CROCUSES

CROCUSES

A COMPLETE GUIDE TO THE GENUS

Jānis Rukšāns

Timber Press
Portland • London

Published in 2010 by Timber Press, Inc.

The Haseltine Building
133 S.W. Second Avenue, Suite 450
Portland, Oregon 97204-3527
www.timberpress.com

2 The Quadrant
135 Salusbury Road
London NW6 6RJ
www.timberpress.co.uk

Printed in China

Library of Congress Cataloging-in-Publication Data
Rukšāns, Jānis.
Crocuses : a complete guide to the genus / Jānis Rukšāns. — 1st ed.
p. cm.
Includes bibliographical references and index.
ISBN 978-1-60469-106-1
1. Crocuses. I. Title.
SB413.C65.R85 2011
635.9'3438—dc22
2010021252
A catalog record for this book is also available from the British Library.

To my love, Guna

Contents

✾ Foreword

It is natural to regard a favorite genus as being rather special and interesting, but in the case of *Crocus* it is undoubtedly true. There are many remarkable aspects of this familiar genus which come as a surprise, even a shock, to those who think of crocuses as consisting of only a few species—with white, blue, and yellow flowers—plus an autumn-blooming species. In fact, the genus consists of more than 80 species exhibiting a fascinating array of characteristics. Even the underground parts of these plants are intriguing, the corms being covered with tunics of varying types. In some cases the corm is so distinct that it is possible to identify the species of a plant in its dormant state.

Crocuses, so beloved of gardeners, display a huge array of subtle features of flower and leaf in their easily visible parts, and even in finer details such as the surface architecture of seeds. Perhaps this is not so unexpected in a genus of this size, but take a look at the more concealed world of their chromosomes and these extraordinary plants take on a rather puzzling dimension. Here we find chromosome numbers ranging from $2n = 6$ rising in increments of two to $2n = 30$ without a break in the sequence; there are species with the same diploid number which are apparently quite unrelated and others with a wide range of genotypes within a species or in closely allied species. Exactly what it all means is somewhat of an enigma, but it is hoped that current DNA studies will shed light on genetic relationships within this entrancing group of plants. In habitat there is also a wide diversity from wet alpine meadows to rocky, sun-baked hillsides, ranging from Portugal eastwards to the Tien Shan.

When invited by Jānis to write a foreword for this book, my first thought was: What do I say if his view of the species and their classification is very different from mine? However I quickly dismissed such thoughts, recognizing that whatever the views expressed this would be an informative volume, full of personal experiences of Jānis's travels to see the plants in the wild and practical information based on many years of cultivating them. Yes, we do not agree on some points, but does this actually matter? The fund of knowledge of any subject is accumulative, added to by each researcher through the ages and is made readily available only through the written word.

My own monograph on *Crocus* was written in 1982, over 140 years after William Herbert was researching the genus, almost 100 years after George Maw's magnificent tome on the genus, and nearly 50 years after that great plantsman E. A. Bowles wrote his rather more gardener-friendly handbook of crocus. Each utilized the previous work and added his own observations and opinions. Thus, another volume is more than welcome and will further enhance this overall fund of knowledge.

As a nurseryman and botanical explorer Jānis Rukšāns has a great deal of experience in the cultivation of bulbous plants of the Near and Middle East and Central

Asia and knows them both as garden plants and in their native environs. His infectious enthusiasm has delighted many audiences and readers in his previous books and in this new book they will similarly be drawn in by his passion for the subject. Yes, some of Jānis's conclusions regarding the classification of *Crocus* are different from mine, but the taxonomic status of a particular plant is of less importance than the information about it—and in this respect there is much to be gleaned in the following pages, for there is a wealth of personal observation for *Crocus* enthusiasts, gardeners, and botanists alike.

BRIAN MATHEW

Preface

MY FIRST BOOK for the international public (written in English) was incredibly successful, and I still receive a lot of letters from readers who are grateful for the information shared in it. When I realized that the experience I had accumulated during 50 years of bulb growing could be of interest and use to other bulb lovers, I decided to share my experience in a more detailed way. And, after thinking long about a possible topic, I chose crocuses.

Why this topic? Crocuses are without doubt one of the most loved plants in our gardens. Several very serious books have been written on these plants. One of the earliest publications to list crocuses as garden plants is *The Names of Herbes* by William Turner in 1548. Before Carl Linnaeus (in 1735) established the taxonomic system that we follow today, many significant works about these very popular and beautiful bulbous plants were published and attempts were made to classify crocuses, but a real revolution was brought about by George Maw. His *A Monograph of the Genus Crocus* (1886) is an excellent source on the history of crocus cultivation. The numerous descriptions and beautiful drawings of the species known to him still haven't lost their value.

The big authority on crocuses was Edward Augustus Bowles whose wonderful book, *A Handbook of Crocus and Colchicum for Gardeners*, was published in 1924 and the revised edition in 1952. It was the main reference on crocuses for several generations of gardeners as well as for me. It has not lost its value as a practical guide to growing crocuses, although since its publication many new species have been discovered.

The real guru of modern crocuses is Brian Mathew whose many years of research on the genus resulted in a marvelous monograph published in 1982. Mathew revised the classification of crocuses, researched the history of the species and their variability in the wild, and presented basic growing recommendations for their garden culture. Given the popularity of crocuses in gardens, it is no wonder that Mathew's monograph has become very rare today and is offered in some bookshops at a price that is several times higher than the original.

Since Mathew's book was published more than a quarter of a century ago, many places where crocuses grow in the wild now are easier to access, travelers have visited almost every corner of the range, and many new taxa have been discovered. The leading experts in these studies are Helmut Kerndorff and Erich Pasche, who both have traveled extensively in Turkey and described numerous new *Crocus* species and subspecies.

It is impossible to name everyone who has contributed to a better understanding of the genus. I have mentioned a few of the most important ones here. Many articles can be found in popular and scientific periodicals on crocuses. An excellent source with detailed descriptions and beautiful drawings of the species is *Curtis's*

Botanical Magazine. Another good resource is the quarterly bulletin of the Alpine Garden Society. Crocus fanciers can join the Society's Crocus group, which annually publishes a bulletin and newsletter at very moderate membership dues.

David Stephens (1998), a National Crocus Collection holder in the United Kingdom, has written: "Every crocus-keen gardener should have a minimum of 30 taxa growing in their gardens for autumn and spring pleasure. If the named cultivars are also grown, you can double or even triple this figure."

So the time for a new manual on crocus growing has come. I decided to do my best to fill this gap with a book that, for the most part, is intended for gardeners. It does not focus on scientific research, but rather is a practical guide to the species, their identification and growing requirements.

Acknowledgments

THIS BOOK HAS been written in a rather difficult time. The world economic crisis forces us to not forget about a budget for everything, and book-publishing is no exception. When I finished the manuscript, I found that it had to be shortened by some 20 percent, and this is the toughest job for every author. So the manuscript is somewhat "drier" than I would have liked, but I hope that you will find my work interesting, informative, and useful for your gardening practices.

First, I want to thank Tom Fischer, Editor-in-Chief at Timber Press, for accepting and promoting my book idea, as well as for his assistance in developing the manuscript.

I have not forgotten the individuals who helped me establish my crocus collection many years ago. Aldonis Vēriņš presented me with my first collection of crocuses and insisted that I write my first monograph on crocuses (in Latvian). In similar ways I got help from Chris Brickell and Brian Mathew (both from the United Kingdom), Eva Petrova (Czech Republic), and many, many others. I especially thank Erich Pasche (Germany) who provided me with copies of his publications and samples from many new taxa described by him and Helmut Kerndorff (also from Germany).

I also want to thank all those who so generously contributed photographs for this book. Their names are listed in the captions accompanying the photos. Two who contributed most to my knowledge about the variability of species are Michael Kammerlander (Germany) and John Lonsdale (United States), but that doesn't mean that the contribution of others is less appreciated. Special thanks to Thomas Huber (Germany) who generously sent me a CD of information from his years of collecting all the available information on crocuses. So many people have sent me rare forms of crocuses that the listing of them all will again require several more pages, but at least a few have to be named: John Fielding, Jim and Jenny Archibald, and Norman Stevens (all from the United Kingdom), Dirk Schnabel and Thomas Huber (Germany), and Marcus Harvey (Australia).

Many thanks go to Henrik Zetterlund of the Gothenburg Botanical Garden. With his help I obtained many rare crocuses for which I had no other source. Our joint trips to mountains searching for new plants are among my best memories.

Of much help to me was my friend, and apprentice and assistant in previous years, Mārtiņš Erminass. He was the first to correct my ugly English (it took a lot of hours and weeks) and his knowledge of Dutch helped me to write the chapter on crocus diseases and pests.

Without the help of my wife, Guna, to whom I dedicate this work, it would not have been possible to spend so many hours each day in the garden with my plants and to travel so often in search of new plants.

To those contributors whose names I have not mentioned here, my apologies. Trust me, you stay in my heart! Many thanks to everyone.

PART I

Crocuses in My Garden

⌘ How It Started

CROCUSES GREET US in very early spring, with the melting snow. In my garden, the earliest crocuses poke their bright flowers through the last snow and continue flowering usually till May. In autumn they are the first and the last blooms in the bulb garden. Only frosts stop their blooming for the winter—a phenomenon unknown in milder climatic zones. Summers go by without crocus blooms, though one species, *Crocus scharojanii*, can be called a summer bloomer as it flowers at the end of July.

Crocuses were one of the first true bulbous plants which attracted my attention in my very early years (besides a few native bulbs growing wild in Latvia) and I fell in love with them at first sight. Today I don't remember exactly where I saw a crocus bloom for the first time. Was it in the rockery of my gardening college where a single corm of a *Crocus chrysanthus* variety bloomed? Or was it in an Estonian garden which I visited together with a group of gardeners from our local horticultural society? There my attention was given to a small bloom of *C. reticulatus* in the garden owner's rockery.

Many years later, when my collection of bulbs was already quite big and I worked as an editor of the Latvian Horticultural Society magazine, my great friend and teacher in bulb growing, Aldonis Vēriņš, proposed that I write a monograph on crocuses. He offered me a good collection of garden cultivars. Actually, it was a complete assortment of crocuses from the Van Tubergen company in the Netherlands.

I also looked for other sources of crocuses, and received considerable help from Chris D. Brickell, at that time director of the Royal Horticultural Society Garden Wisley, who sent me seeds of the newest taxa. Michael Hoog sent me E. A. Bowles's *Handbook of Crocus and Colchicum for Gardeners*. This was not only my first crocus book but also a sort of textbook in English. I read it many times from the first to the last page, I read it everywhere and anytime—in the evenings before going to bed, during lunch breaks, on the bus, train, or plane as I traveled—and slowly I began to understand more and more.

Through seed exchanges at botanic gardens I received plenty of seed, but my biggest help came from Eva Petrova, supervisor of the bulb collection at the Institute of Ornamental Horticulture in Pruhonice near Prague (then Czechoslovakia). She kindly supplied me with many wild species and even today I grow the true wild *Crocus minimus* which I received from her.

My study to that point resulted in a 175-page monograph on crocuses, written in 1981 in Latvian. At that time my collection consisted of approximately 200 various crocuses—mostly garden cultivars, fewer wild species.

Later on, my collection would increase and decrease, mainly because of rodent activities, though the vagaries of weather also exerted considerable influence. It wasn't until I built my first greenhouse that I was able to grow most of the autumn-blooming crocuses. None of the these species, which overwinter with more or less developed leaves, are suitable for growing outdoors in Latvia, where winters are too changeable and harsh for them. Now that I have several unheated poly-tunnels, I can grow with some care almost all the crocus species, with the exception of a few species from the southernmost areas in the wild which need much more light in winter than what they can get here. Some German and English enthusiasts have provided their greenhouses with additional lighting and heating equipment, so that almost all the known species are currently being grown somewhere. I have most of them but not all, and many new species are still waiting to be discovered in the wild.

At present, there are about 700 samples of crocuses growing in my collection. The number changes all the time as some plants die, others turn out to be wrongly named, and still other new plants are added to the collection. Some of the species in my collection are represented by several dozen plants, while others by only one or two corms.

Growing in an Open Garden

Crocuses grow in a very wide area stretching from northwestern Africa through southern and central Europe and the Middle East to as far as China. They are very variable in cultural requirements: some species come from forests, others from deserts, with intermediates between them. This diversity makes it impossible to give any universal recommendation of their cultivation. Therefore specific requirements will be given when discussing the individual species.

Naturally, some universal tips apply to most crocuses. George Maw (1886) went so far as to write that advice-giving for such an easy garden plant as crocus was completely unnecessary. But, in reality, things are not that simple. Some species are easy to grow and thus widely grown in almost every garden with bulbous plants; other species, usually described as "rare," are more difficult to grow. It is possible to have occasional spectacular success with those "rarities," but just when you start to applaud yourself for this achievement, they perish. Some crocuses survive but never increase, some set seed very well, but some never flower for you. In many cases these problems are caused by a lack of understanding the conditions in which the plant grows in the wild.

Until the 1990s I grew all of my crocuses in the open garden. This seriously limited the number of species I could grow. Crocuses generally like light soils, although in the wild they often grow in rather heavy clay. What is important is the whole complex of conditions. The mountain slopes where many wild crocuses grow are stony, have excellent drainage, and are dry in summer, while conditions in most gardens are very different, making drainage extremely important. I know of growers who have tried to imitate the soil structure found in the wild in their own gardens but eventually failed. Only after they reverted to very porous and well-drained mixes did these growers obtain acceptable results.

For purposes of cultivation, I divide crocuses into three basic groups:

Group A. Crocuses that grow in quite harsh conditions with very hot and dry summers. These crocuses do not grow well outdoors in areas with rainy summers. Many of them can be grown outside but, to ensure a hot and dry summer, will need annual lifting and storing during the summer in a bulb shed. Typical examples are species from Syria, Israel, and Jordan.

Group B. Crocuses that do not require extremely dry summers and can also tolerate occasional summer rains. These crocuses are very easy to grow in most gardens. They don't like being out of the soil too long, especially if the weather is very hot, and they can be left in the soil for two or three years without replanting. Typical examples are the large Dutch crocuses derived from *Crocus vernus* and the cultivars of the so-called *C. chrysanthus* group.

Group C. Crocuses that grow in moister conditions in the wild and can seriously suffer if kept out of the soil for too long. These crocuses grow in conditions opposite those of the first group. And, once again, the group varies greatly, from *Crocus scardicus* and *C. pelistericus*, which in the wild grow in sites that never dry out in summer, to *C. heuffelianus*, which you can lift and store in the bulb shed for a long time without special precautions. Of course, species of this group grow best when left in the soil undisturbed until the clumps become too congested and need dividing. Some even need special attention in the form of an additional watering.

In several instances I have seen crocuses growing near streams in soil that is very damp, though this is the case only in early spring. Crocuses need plenty of water during the period of growth, and in early spring even prefer that their root tips stay in water. The lack of water during the vegetation period can seriously decrease the size of the lifted corms and their numbers.

Although crocuses need a lot of water they don't like waterlogged sites. Good drainage is of paramount importance. For this reason, I plant crocuses on raised beds and on slopes (Plate 1). Dutch growers do likewise, only beds are wider, planted by machine (Plate 2). Proper drainage is the key to growing bulbs, though slopes and soil structure are not the only ways to ensure it. I also plant all my crocuses (and other bulbs as well) in sand beds. I strew a layer of coarse sand on the bottom of each row before placing the corm on top and covering it with more coarse sand. In this way the corms stay in pure sand during their entire growth period.

This planting method gives me many advantages. First, it provides additional drainage. Second, it prevents the planted corms from direct contact with the soil. The third advantage, however, is the most important. The sand usually is very light in color and contrasts well with the surrounding dark soil, so that when it is time to lift the corms out, I simply take the light sand together with all the corms and scatter it over a sieve. The sand falls through and the corms stay on the sieve. This method works especially well when digging up species that form small corms or when digging up late in the season when the leaves have withered away and it is quite difficult to find the corms in the soil.

Tony Goode (2003) tried a special type of sand bed, which he calls a pseudo-sand bed:

> I have a raised bed with about 15 cm of sand over a layer of old potting compost. The corms are planted in the sand but will root through into the richer mix below. The idea was to give a warmer, drier rest to the corms while maintaining access to nutrients. *Crocus goulimyi*, *C. kotschyanus*, *C. pulchellus*, *C. hadriaticus*, and *C. flavus* have persisted while *C. imperati* has not.

Crocuses need nutrients just like any other plants. The best fertilizer is bone meal (up to 100 grams per square meter). Unfortunately, it is not available here any more, so I have replaced it with a slow-release granulated complex fertilizer with all the micro-elements. I apply this fertilizer at the rate of 70 grams per square meter approximately two weeks prior to rototilling the soil before planting. Such complex fertilizers contain nitrogen as well, an element that is very important for leaf growth. Although all gardening manuals usually caution against nitrogen-containing fertilizers, they are not dangerous when used at the proper dosage. Every plant needs nitrogen.

Once, in very early spring right after the snow had melted, I mistakenly sprinkled my beds of *Crocus vernus* cultivars with ammonium nitrate at the rate of 200 grams per square meter. The crocuses didn't die, but they did produce 70-cm long leaves which didn't wilt until the end of June. The crop of corms was normal and they bloomed normally the next spring. Usually I spread my crocus beds in early spring with ammonium nitrate at the rate of 40 grams per square meter, and repeat the process three weeks later at the same rate but using a complete fertilizer with micro-elements.

It is very important here to plant crocuses early. The autumn-blooming species must be planted first, as some of the earliest blooming among them will start to flower in August while still resting in boxes in the bulb shed. Thus the best time for planting most crocuses is late summer: you get excellent rooting and, the following season, abundant flowering. Only species from the previously described Group C must be replanted as soon as possible after their corms are lifted.

If you plant corms in September, this decreases the blooming performance of the next season. Late-planted crocuses form weaker root systems before frosts set in and the following spring their flowers will not be able reach the surface of the soil, so they will start blooming underground. Planting crocuses in October is much too late here and will seriously affect not only the flowering but also the crop of corms. In any climate, it is important to plant corms at the right time of year. In 1809 Adrian Haworth wrote that crocus corms could be kept out of soil until September if necessary, but the longer they stay unplanted, the weaker they became and they bloomed later.

Several small tips can help you to preserve your collection. The first is to divide your stock of a particular plant into two groups, even if you have only two corms, and plant them in separate beds. This strategy helps save your treasures in case rodents attack one of your beds.

Another tip aims to prevent the accidental mixing up of stocks. The external appearance of bulbs is different in many species, and for some species, such as *Crocus fleischeri* and *C. goulimyi*, you can identify the species with certainty by seeing only the corm. To avoid mixing up corms, especially when the stocks are small and the labels get lost, it is essential to plant crocuses with different types of corm tunics beside each other. For example, plant crocus stock with a reticulate tunic, the next with an annulate tunic, followed by a one with a fibrous tunic, and so on. Such a planting pattern greatly helps to prevent the mixing up of stock, because you can immediately see where one stock ends and another begins simply by looking at the corms.

The planting depth depends on the purpose for which you are growing crocuses. If you want to replant them infrequently, plant them deeper; they will form large corms and the natural rate of increase will be much slower. If you want to increase the stock, plant the corms shallower; they will form smaller corms but the natural rate of increase will be faster. Zinaida Artjushenko (Russia) experimented with the planting depth and found that, three years after planting, corms planted 10 cm deep formed 18 corms of 3.5 cm in diameter each but corms planted only 5 cm deep formed 32 corms of 2 cm in diameter.

I usually plant crocuses approximately 5 cm deep but I cover my beds with a 5-cm layer of mulch—I use peat moss—so the corms actually are at a depth of 10 cm. E. A. Bowles recommended a depth of 10 to 15 cm, noting that *Crocus flavus* and *C. speciosus* can be planted even deeper. In the wild I have found crocus corms at depths from 4 to 23 cm below the surface. In my old field the maximum depth from which I once lifted two corms of *Crocus* 'Jeanne d'Arc' was 47(!) cm. Both corms bloomed and were very large in size.

The first job after planting bulbs is to cover the beds with a 5-cm layer of mulch. Mulch keeps the soil from becoming too compact and the moisture from evaporating, and is very essential in protecting corms against excessive frost. As soon as the beds are covered I water them. Watering encourages the roots to start growing. Good roots are needed for the corms to overwinter well, and they also boost plant hardiness.

If crocuses have suffered from frost, don't rush to dig them up. Sometimes only the roots and the base of the corm—the parts that are the tenderest to frost—are damaged. The young buds covered by old sheathing leaves are hardier and can survive and form a small replacement corm from nutrients that have remained in the old corm. The next season there will be a grasslike growth and in two to three years the new corm will reach flowering size. This happened to me once: despite a terrible black frost in 1972 in which the temperature during several nights in January dropped to –27°C and the ground was completely snowless, my beds with frost-damaged *Crocus speciosus* cultivars again bloomed nicely in the autumn of 1975.

The best time for lifting bulbs is when the leaves are completely dry. Corms that are lifted too early suffer more from drying out (mummification) during storage in the bulb shed. Furthermore, it is also more difficult to clean the young corm from the remains of the old corm. It is very easy to damage the basal plate of the young corm when removing the old corm if the corm is harvested too early. The damaged young corm almost always dies.

Only a few species like *Crocus scharojanii*, *C. scardicus*, and *C. pelistericus*, which have nearly non-stop roots, must be replanted before the old leaves die. *Crocus veluchensis* forms new roots before the leaves have died, therefore it is better to lift it before the new roots have started to form. After several cases of serious losses, I now postpone the cleaning of harvested corms. I wait until it is time to dispatch the corms to my customers or until it is time to plant the corms before removing the old roots and the remnants of the old corms. This late cleaning help protects the corms from mummification. E. A. Bowles wrote that "the withered portion of last year's corm . . . (can be removed) . . . from the base if it comes off easily and without use of force."

Sometimes, if it is absolutely necessary, crocuses can be lifted before the foliage has died, but then it is best to keep the corms in boxes covered with soil in some shade and to not water them. Watering such corms can only cause rot.

How long can crocuses be grown in the same place without replanting? Many authors state that the maximum is four years, noting that in the fourth year the flowering decreases significantly. Both George Maw and E. A. Bowles recommended annual replanting when not naturalizing bulbs in grass, only allowing two-year-long cultivation in exceptional cases. At the National Botanic Garden of Latvia, Velta Amatniece (1975) found that the number of corms of *Crocus chrysanthus* 'Saturnus' grown in a two-year cycle decreases 6.4 times compared with corms that are replanted annually. My own experiments with *C. tommasinianus* 'Ruby Giant' showed that after two years of uninterrupted growth each corm that I planted yielded 6 to 14 new corms (on average 8.6), all of which were small. The same cultivar if replanted every year yielded on average 4 new corms per planted corm, all of flowering size; thus in two years one corm increased to 16 corms.

Another problem that arises when crocuses are not replanted annually is that many of them self-seed on my nursery beds. Because the seedlings of cultivars differ from the parents, it is very easy to get mixed stocks. In my old crocus field all the pathways between the new beds are packed with *Crocus heuffelianus*, *C. abantensis*, *C. chrysanthus*, and some other seedlings.

The best temperature for storing crocus corms after lifting is around 17°C. Higher temperatures increase the risk of mummification and delay flowering in spring (according to experiments by A. Kolcova at the Nikitsky Botanical Gardens, Yalta, Crimea, Ukraine). At lower temperatures, the risk of the corms forming roots while still in the bulb shed increases, and that makes planting more difficult. Sometimes long shoots are also formed; they are very brittle and break easily during planting.

Crocuses are very popular for naturalization but in my area only a few species can withstand the competition of grass (Plate 3). I had some success naturalizing crocuses when I planted them in slightly shaded spots near trees where the grass wasn't too dense. Actually, I was successful with only two species—*Crocus heuffelianus* and *C. tommasinianus*. Both self-seed and abundantly flower year after year. A third species that I have successfully grown for a quite long time in scanty grassland is *C. herbertii*. It likes somewhat moister conditions and spreads slowly by stolons. E. A. Bowles recommended removing a 15-cm layer of the topsoil before laying the corms 2.5 cm apart from each other. The results depend on the quality of the grass. The denser the grass, the poorer the crocuses. Of course, the first mowing should be done only when the crocus leaves have died down.

My wife plants crocuses between clumps of herbaceous peonies and hostas with success. Because the growing season of peonies and hostas starts here in the second half of May, the crocuses have enough time to mature. At the Gothenburg Botanical Garden in Sweden, crocuses and corydalis are planted together (Plate 4).

Crocuses can be planted on south-facing rock garden slopes. It is better to place corms in smaller groups filling in the empty spots. Most groundcovers are tough competitors for crocuses, which quickly disappear.

✻ Growing in Pots and Containers

In *Buried Treasures* (2007) I wrote that I didn't like growing bulbs in pots, but if you were to visit my nursery today, you would see thousands of pots. Every year I learn something new and I gain new experience, all of which changes my growing methods.

For many years I grew a number of bulbs in greenhouses but not in pots. I made raised beds and planted the bulbs directly in an artificial soil mix. I still think that bulbs grow better and reach a bigger size when they are in the ground rather than in the limited space of a pot, but a number of practical reasons have caused me to change my growing methods, especially when small stocks are concerned. One disadvantage of growing bulbs in pots is that they need more attention and care than when they are grown in beds. At the same time pots have a lot of advantages, the most important of which is that plants are separated from each other and this prevents the spreading of diseases. It is easier to keep your stocks clean, especially when they are small, and when a problem does arise, you destroy or set aside potted plants easier than you can non-potted ones.

My first attempts at pot culture started with seedlings. My stock of many crocuses is very small, so the number of harvested seeds is likewise small. If the seeds are sown in large boxes, it is much harder to keep the stock clean. This problem was solved by planting seeds in pots. Only one name, or plant taxon, is sown in each pot and in such a way that I can be 100 percent sure that the resulting seedlings are only of the sown stock.

Next, I started using pots for my smaller stocks of crocus plants. Although clay pots are generally accepted as being more suitable for bulb growing, their round shape wastes plenty of unused space between them (Plate 7). In addition, moist clay pots split easily in winter frosts. Therefore one year later I replaced the clay pots with square plastic ones (Plate 5). Fortunately for me, I was able to find very deep plastic pots of 20-cm depth (the so-called rose-pots).

These plastic pots have large holes at the bottom, allowing for good drainage, but I always put a handful of coarse sand in the bottom of a pot. The sand must be moist or it will fall out through the drainage holes. The sand I use contains a little bit of clay so that when it becomes dry, it hardens slightly and cannot fall out of the drainage holes if the pots are moved.

Any soil mix for crocuses must be very porous. I combine two parts coarse sand, one part peat moss, and one part loam, mixing thoroughly with a rotavator. I add dolomite dust to ensure that the pH level is around 6.5. I also add several handfuls of a slow-release fertilizer that includes all the necessary microelements and the basic elements (N–P–K around 10–10–20). This mix is sufficiently porous to allow water to drain and, at the same time, retains moisture well, too. Some gardeners in other countries add small stone chips or perlite to the mix, making it still lighter.

Crocus expert Tony Goode (United Kingdom) wrote:

> [J]ust as important as the mix is how you manage watering and feeding. There is no substitute for personal experience; each grower should develop a watering strategy to suit their own compost and growing conditions. The compost should be moisture-retentive but free draining and contain enough nutrients for the long growing season. I use a mix of roughly 50% John Innes No 3, 45% sharp grit or gritty sand, 5% perlite. To this I add bonemeal for a slow-release feed during the growing season supplemented by a high-potash feed in late winter. Following Ian Young's approach I add a small quantity of sulfate of potash to each pot and water it in. This high-potash feed boosts the plants at the time when the new corm containing next season's flower buds is developing.

Helmut Kerndorff and Erich Pasche (Germany) use a mix of two thirds of greywacke (a kind of sandstone) mixed with sharp sand (50:50) and one third standard soil (consisting mainly of peat with an addition of clay minerals). To this they add 3 kilograms of bonemeal and 100 grams of microelement fertilizer (sold under the trademarked name of Radigen) per cubic meter of soil. After placing the pots on greenhouse benches (they use square plastic pots), they cover them with a 3-cm layer of pine needles which keeps the soil surface clean from moss and cool during hot summer days, and prevents soil particles from splashing on the flowers during watering. It smothers weeds and minimizes the impact of frost.

In my greenhouses, I fill the pots with the soil mix I've prepared, then press it down with a special tool, freeing up the top 6 to 7 cm of the pot. I add a thin layer of coarse sand on which I lay the crocus corms, then cover the corms with more coarse sand and fill the pot to the top with the soil mix. Using my special tool, I again press down on the mix, freeing up 1.5 to 2 cm at the top of the pot, which I fill up with small stone chips. I follow this same procedure when sowing seeds, only this time there is more soil in the pot because the seeds are covered by a thinner layer of substrate.

Such pots need watering at least once a week in normal temperatures, twice as frequently in hot spells (Plate 6). Plastic pots do have one advantage—they dry out slower than the clay pots when plunged in sand. At the same time, it is slightly more difficult to control the moisture level, so I sometimes use a moisture meter (available at any garden center). I pay attention to the moisture level at the bottom of the pot, making sure the indicator shows "moist" during the active growth period.

Some growers use leaner soil mixes, but then daily watering and regular fertilizing are necessary. The leanest mix that I've heard of is made from 1.5 parts gravel, an equal amount of pumice, and one part each of perlite, sand, and dry clay (Michael Kammerlander, report at the First Czech International Rock Garden Conference, 2007). Pots then are fertilized up to five times during the season, and bulbs repotted every year.

I try to repot all my crocuses annually. As a rule I replant them immediately after harvesting. Only crocuses listed in my catalog go to the bulb shed.

During the active growth period, I spray my plants weekly with a weak (0.2 percent) solution of a water-soluble complex fertilizer containing all the microelements and the basic elements (crystalline type). At the start of the season, I use fertilizer with an equal ratio of nitrogen to potash, but in the second half of the season, I switch to something with less nitrogen and more potash. With such care, the crop of corms is good, probably not always as bountiful as in the open garden or on open beds under cover, but they always grow and increase well.

Helmut Kerndorff and Erich Pasche recommend using granulated fertilizers with 30 percent potassium oxide (K_2O), 10 percent magnesium oxide (MgO), and 18 percent sulfur (S) at the dose of 30 grams per square meter plus a weekly feeding with an 8–12–24 N–P–K solution at the dose of 2 grams per liter of watering.

In my nursery, crocus pots are placed according to the plant's growing requirements. First are the autumn-bloomers that need wet conditions, then the spring-bloomers that need wet (Group C). Next are the autumn-bloomers that are tolerant of moisture, followed by the spring-bloomers that tolerate moisture (Group B). Last are the crocuses that need a hot and dry summer rest (Group A). In the summer, I move the pots with Group B and Group C species outdoors. Pots with Group A species stay in the greenhouse all summer, although the overhead and side windows are opened all the time to lower the temperature.

The question arises about introduction of crocuses from the opposite hemisphere. I have no experience in this area as all crocuses are plants from the Northern Hemisphere. Recently, however, I wanted to reintroduce into the north some rare forms grown by my Australian friend Marcus Harvey. He recommended the following:

> Crocuses are quite easy to acclimatize. Store the spring-flowerers in damp perlite in a fridge for six to nine weeks and you will have them showing a big bud. Plant them slightly more shallow than usual and water them; they should show immediate signs of bud burst and quick growth. Keep them growing through to early summer and then give a dry period during late summer to early autumn and treat as you would your normal stock. Usually you will get smaller corms or clusters of corms on the old corm because of interrupted or late growth but this will improve the next year.
>
> As for the autumn-flowerers you just have to plant now and keep them as stable at autumnlike conditions as you can. Here in Tasmania the weather is not as extreme as yours, so I can plant autumn-flowering bulbs and they emerge on your Northern Hemisphere times and continue to grow through our spring and part of the summer (under shade) until I withdraw water or the weather becomes too warm in January or February. Here now is September and I still have a few of your crocus trying to keep on growing (*Crocus tommasinianus* and *C. vernus*) and should really be stopping them by storing the pots in a water-free zone.

ꕥ Winter in the Greenhouse

To protect my plantings in greenhouses, I use 5-cm thick sheets of glass-wool insulation, the same material used in the building industry for heat insulation (Plate 8). These sheets are very lightweight and long-lasting, and ensure excellent protection. The elevated bulb beds in my greenhouses are now built in accordance with the size of my pots and boxes and the size of the glass-wool sheets. Not all companies produce them in the same size, so know the size of your beds when you go to buy glass-wool sheets. Rock wool can also be used, but it is brittle and doesn't last long.

Helmut Kerndorff and Erich Pasche use a 3-cm layer of pine needles but winters in Germany are much milder than those in Latvia. The advantage is that pine needles can stay on the pots all the time and they can be easily taken off.

In a small greenhouse, a fan heater is the best way to keep plants nearly frost-free in winter. The heater not only keeps the air moving but also warms the air. When I had only one medium-size greenhouse I used a large hot air blower during the coldest nights, but the electrical bills I received were huge, so after a few seasons I decided that it would be wiser to not grow the tenderest of species than to go bankrupt.

Other problems in greenhouse culture are the lack of light and the excessive air humidity during the dullest winter months. Such conditions are very favorable for *Botrytis* and *Fusarium nivale* infections. For that reason it is very important to remove all the dead flowers from the autumn-blooming crocuses. The old flowers are the first to get infected and through them the infection reaches the shoots and in the end the corm. Therefore, using fans to get the air moving continuously is very important, especially on colder days when the windows are closed. Even in spring, after each watering, fans must be used so that the leaves dry quicker. One beneficial side effect is that the transpiration rate intensifies as well, which in turn increases the absorption of water by the crocus roots and, along with the water, nutrients are absorbed.

Many autumn-blooming crocus species form leaves simultaneously with the flowers or immediately after the end of flowering in autumn. Overwintering such crocuses is more difficult. I found that the plants can be covered with sheets of insulation, as mentioned previously, but before they are covered, the soil must be allowed to freeze slightly.

The leaves of plants overwintered in this way do quite well, although sometimes the foliage is damaged at the point where it bends and is in direct contact with the covering sheets. Come springtime, these spots yellow and dry within a week or so after the cover is removed. Fortunately, crocus leaves grow from the base and a new chlorophyll-containing part grows up below the dead zone. The same happens in the wild when animals graze off the leaf tops; the remaining part of the leaf elongates and provides the new corm with nutrients. The corm crop, however, is smaller.

Crocuses that flower in midwinter pose a problem. They are in full bloom just when the covering is needed. I am not sure what is the best thing to do—to remove the flowers or to leave them under the cover. Up to now I have left them until spring, but then they rarely are attractive after the cover is removed. Usually they have suffered from some frost damage.

Growing from Seeds

It is very important to grow crocus from seed. Interestingly, in difficult seasons, when mature corms perish, it is the seedlings that survive. Many times I have not lost rare species thanks to the seedlings I grew. Other times only the seeds that had been sown the previous autumn and had not yet started to germinate survived a difficult winter. For this reason, I recommend that you always collect seeds, even if you will not actually sow them. You can put them to use in a seed exchange.

Propagating crocuses from seed has several advantages. Seed-raised progeny usually are free from viruses. Raising seedlings also helps to increase the stock faster than waiting for the stock to build up vegetatively, particularly in the case of species that are slow to increase vegetatively or do not do so at all.

Even in an ideal situation, where a grower starts with a good number of seeds which germinate well, the number of seedlings can decrease. During the years before the seedlings start to bloom, plants may be lost because of pests, diseases, your own mistakes, or simple misfortunes and other circumstances that work against you. Eventually, however, you will end up with a few flowering plants, which, thanks to natural selection, are adapted to your growing conditions. These plants retain the genes that ensure survival for coming generations.

An additional advantage (or sometimes disadvantage) of growing crocuses from seeds is the appearance of hybrids among seedlings. Most *Crocus chrysanthus* varieties raised in my nursery originated as unintentional hybrids. I have never crossed them intentionally. Instead, that job has been performed by bees. I only collected seeds and sowed them in favorable conditions. What a myriad of beautiful colors came up, when they started to flower! But if you want to be certain that your seedlings are true to name, you have to isolate the flowers and pollinate them by hand.

When you decide to raise your stocks from seeds you collect, you have to pollinate them yourself. Crocuses generally bloom at the time of year when weather conditions are very changeable, and in some years there are very few sunny days when the flowers are open for pollination. If your crocuses grow in pots, you can bring them inside where the dry and warm room air will quickly open the flowers. In these conditions you will need to pick the right moment to pollinate your plants before the stigma dries out. In recent years I have regularly hand-pollinated the very early blooming Central Asian species—*Crocus michelsonii*, *C. alatavicus*, and *C. korolkowii*—and gotten good seed crops. After pollinating the plants, the pots can be returned to the greenhouse. In hot, sunny weather the anthers open very early and pollen falls out, so it is best to use such plants on the day the flowers open. Sometimes I have succeeded in collecting pollen on a brush from a flower that was only half open.

The stigma retains its ability to fertilize for a very long time; there are reports about successful pollination even of wilted flowers at the very end of blooming. Later in the season, when bees start to come around, another problem arises—how to prevent an undesirable cross-pollination. You will need to cover the pots of pollinated flowers with muslin bags or tight wire mesh to keep insects away from the flowers. I don't use any covering on my plants because, in most cases, I'm happy when unexpected hybrids appear among my seedlings now and then.

To transfer pollen from one plant to another, I use a very thin brush made from squirrel or camel hair (Plate 9). Another way to do this is to remove the anther from the flower using forceps, then brush it against the stigma. This method may damage the flower tepals.

Recently, while rereading *My Garden in Spring* (Bowles 1914) I learned another method of pollination:

> When you see your crocuses wide open in flower sally forth with the stick of sealing-wax or the amber mouthpiece of an old pipe in your hand. . . . Rub whichever of the two unusual accompaniments of a garden stroll you have chosen, on your coat-sleeve if it be woolen, and hold the rubbed portion as soon as possible after ceasing rubbing near the anthers of an open Crocus, and you will find the electricity thereby generated will cause the pollen grains to fly up on the electrified object, and, what is more, to stick there, but so lightly that directly they are rubbed against the stigma of another Crocus they will leave the amber and be left where you, and Nature before you, intended them to be.

And then I remembered how that when I was a child my father demonstrated the effect of electricity. He would rub his amber penholder or plastic hairbrush against his jacket, and then small pieces of paper would fly up and stick to the penholder. I immediately tried this method on autumn crocuses in my greenhouse and to my surprise found that it worked. In fact, this technique is much better for transferring pollen than either of the other mentioned techniques. When using an amber tool or sealing wax (the latter not tried by me), pollen can be easily wiped clean between different crosses. And with a brush, some pollen grains tend to stay in the fine hairs and do not always stick to the stigma. Then, when using the same brush for the next cross, those grains can be transferred, thus muddling the crosses.

Crocus seed capsules emerge from the soil at the end of vegetation. Only seedpods of a few species (for example, *Crocus caspius*, *C. korolkowii*) remain underground. Some (for example, *C. pelistericus*, *C. scardicus*) are pushed up on long stalks and ripen very late in the season. It is easier to collect seeds right before the capsules are split. The stage of ripening can be checked by slightly squeezing a capsule between the fingers—if it is hard, seeds are ready to be harvested. It is better to store them in small boxes in a shaded, warm, and dry place where the capsules will soon open. I prefer to wait to gather seed until the moment the capsules start to split—then you can open them and allow the seeds to dry up a little (Plate 10).

What is the best time for sowing seeds? One can find plenty of recommendations. In recent years I have sown the seeds as soon as possible after harvesting. This has given me the best germination rate, although in a few seasons I've had problems with the seeds germinating way too early. When early sown seeds start to germinate in the autumn, the seedlings they produce often die in the winter that follows. For species that overwinter with leaves, autumn germination is very common and causes no problems. Seed of such species is best sown as soon as possible after it is harvested. In most cases, losses from seeds germinating too early are smaller than losses from delayed germination when seeds are sown late.

Bowles (1952) recommended sowing seeds no later than the first week of September. Older seeds germinate within several years but that depends on the species. Maw (1886) wrote of his experiment with four species. When he sowed 40 one-year-old seeds per species, the number of seeds that germinated the first year was 25 for *Crocus imperati*, 2 for *C. vernus*, 3 for *C. flavus*, and none for *C. versicolor*. In the second year, 9, 22, 1, and 13 seeds germinated, respectively.

Usually I have had the first flowers on seedlings in the third year after sowing, but the blooming starts from the fourth year onwards. David Stephens reports that he got the first flower of *Crocus gilanicus* in a pot of 20 seedlings in the autumn of the second year after sowing (27 months later) and the first flower of *C. gargaricus* 18 months after sowing. With me, seeds from a cross between *C. abantensis* and *C. ancyrensis* sown in October 2007 gave their first flowering in March 2009—17 months after sowing. I replanted the small seedlings in the third year after sowing.

I also sow my crocus seeds mostly in boxes or pots. A few species which are excellent growers outside and which set plenty of seed I sow in garden beds. Among these select few are *Crocus abantensis*, *C. chrysanthus*, *C. heuffelianus*, *C. korolkowii*, *C. malyi*, *C. veluchensis* (Plate 11), and *C. versicolor*.

I would like to end this chapter with an extract from E. A. Bowles (1952) (I realize I quote him frequently, but no one can say it better than he can):

> The first three barren years soon pass away, and then it is good to stand before the seed bed on a sunny morning and see the rows of open blossoms with here and there one that is unlike its brethren and perhaps anything else we have seen before. A pure white seedling of *Crocus sieberi* rewarded me for thirty years of patience, and I still hope for further pleasant surprises.

Pests and Diseases of Crocuses

Like any other garden plants, crocuses have many pests and diseases that seriously damage or even destroy the plants. Not all of these, however, are widespread. In this chapter we will look at the pests and diseases commonly encountered in gardens. For professional growers and others wanting more details, an excellent resource on the various diseases and pathogens of bulbous plants has been published by a Dutch organization (Laboratorium voor Bloembollenonderzoek 1995). In it, one chapter is devoted to crocuses, and each malady is illustrated with colored illustrations to help readers identify problems with their plants.

Pests

The number one enemy of all bulbs, and especially crocuses, is the various kinds of **rodents**. Whether mouse, vole, or rat, all have an insatiable appetite for the rarest and most expensive gems. In my collection rodents once completely destroyed a beautiful form of *Crocus biflorus* subsp. *adamii* that I had collected near Tbilisi in Georgia and carefully multiplied over the years from a few corms to more than 200 specimens. Not one corm survived this attack. On another occasion, rodents almost completely destroyed my similarly sized stock of *C. tauricus* from Crimea.

At my nursery, the most dangerous rodents are the water rats (or water voles) that live in the ponds and bogs surrounding my nursery. These pests rarely leave their holes in broad daylight and are very cautious when presented with new kinds of food (such as poisons). The best way to protect a garden from rodents is to clean up the surrounding areas, thinning out thick shrubs and regularly cutting the grass which provides hiding places for the rodents from their natural enemies, various birds of prey. Now that I am aware of the most dangerous spots in my garden, I avoid planting bulbs there, but in the early years I lost enormous numbers of bulbs because of water rats. One time I lost 2000 (!) corms of *Crocus scharojanii* to these pests. Not one corm was left.

There are a couple of more or less effective controls in the endless battle with rodents. One natural way is to keep a cat or two (I have four), but unfortunately cats seem to prefer birds over mice. Other natural predators of rodents are owls and foxes. I have placed owl houses in the large trees surrounding my nursery to encourage owls, and to protect the foxes that live in the nearby forest, it is strictly forbidden to hunt them on my property. I don't like traps because small birds are sometimes caught in them or even a cat's paw, but poisons can be very effective. On every 100 square meters in my bulb fields is a short tube in which I regularly place granules of rodent poison. I do the same in my greenhouses. The size of the tube prevents birds from entering and eating the poison.

Another way to preserve at least part of a bulb collection is to split all the stocks into two groups, and plant them at some distance apart. As additional protection, I grow a few corms of each bulb, even the easiest and most common species, in pots. Rodents don't like rummaging in the small stones that cover the soil, and corms, in pots.

Moles, which are protected in some parts of Europe, assist rodents by making holes in the ground that make it easier to access the bulbs. The only sure way of getting rid of moles is to trap them. Traps are most effective in early spring, when the moles are searching for a mate and lose their usual vigilance, and in autumn, when the moles are gathering as much food as possible for winter and thus run swiftly in their tunnels. Moles play both positive and negative roles in the garden. On the one hand, they destroy plantings, dig holes and molehills, and clear the way for rodents. On the other hand, they eat grubs, another pest of bulbs.

Grubs live in the soil and damage plant roots and corms. In my garden, maybug and cockchafer grubs are serious pests in certain years. There are seasons when these pests increase in enormous numbers, causing damage to the roots of young trees in forest plantings and invading gardens as well. A few years ago while harvesting crocuses in the garden I found on average five big maybug grubs per each meter of a crocus row. Many crocus corms were seriously damaged or completely eaten. That season I replanted my pots with *Crocus scharojanii*, which had been growing for four years since the last replanting, and I was shocked to find in each pot three to five grubs and very few crocus corms. The development cycle of maybugs and wireworms, which like crocuses as much as maybugs, is four to five years. During this time, the insects live in the soil, eating roots and corms. They only come out of the soil at the end of the cycle to lay eggs. The best method of control is annual replanting and mechanical killing of all grubs while preparing soil for the next planting.

Ants can be unpleasant pests if you collect seeds. I have noticed them in greenhouses where they sometimes eat the sides of seed capsules, revealing the seeds and consuming the caruncle of unripe seeds. Some seeds stop their development after such damage. Sometimes ants carry seeds a great distance away, and they can mix seed stocks that are kept in small boxes in a greenhouse to dry.

Several **nematodes** are pests of crocuses. If the crocuses are infected by crocus corm nematode (*Aphelenchoides subtenuis*) or potato nematode (*Ditylenchus destructor*—very rarely), the first symptoms can be discerned in the bulb shed. Root tips at the ring where they come out appear to be bloated and are dirty white or pinkish in color. First symptoms appear in the seventh week after lifting. Gradually the corm tissue darkens and becomes dark brown. Cutting a corm in two parallel to the basal plate, sometimes reveals a round or star-shaped pattern. Infected corms do not form roots, shoots seem compressed and sometimes they bloom a few days before the healthy plants. The tips of leaves are whitish or brownish and they die down earlier. Infected mother corms usually do not rot and may even form a few small replacement corms on their tops, but these usually are infected, too. The best thing to do is to destroy all the suspicious corms. Dutch growers use hot-water treatment at 43.5°C after 10 days of pre-treatment at 25 to 30°C. Corms for sale are treated in hot water for one hour, those to be replanted, for four hours.

In some places **birds** can cause serious damage to plantings. Unrivalled demolishers are crows, jackdaws, and magpies which in early spring especially like the bright-colored flowers of crocus. Nets spread over garden beds offer some protection from these creatures. **Hares** and **rabbits** can eat the green leaves of crocuses; rabbits sometimes dig up the corms, too. Similar damage can be done by **deer**, which in early spring love the fresh green salad from crocus leaves very much. In the United States a serious enemy is the gray **squirrel**, which can dig up all planted crocus corms. In some regions crocus growing is possible only if the corms are placed in wire-net cages that keep out squirrels and underground gnawing pests. The European red squirrel has never damaged my crocus plantings, although it lives in the vicinity.

Snails and slugs can be a big problem in milder climates, such as the United Kingdom. They have never attacked my crocus plantings in the open field or in greenhouses here in Latvia, so I am inexperienced in protecting crocuses from these pests. As far as I know, mechanical collecting and some chemicals are widely used elsewhere to control these pests.

Bacterial and Fungal Diseases

Only one **bacterial disease** affects crocuses, and it does so rarely. *Pseudomonas gladioli* causes cavity formation on corms. The cavities are coated in a glossy hard substance. This disease usually affects only corms that are damaged by bugs or mites before they are planted and does not affect undamaged corms. It is most dangerous in freshly broken ground and can be eliminated by killing the pests.

Mildew is caused by *Penicillium hirsutum* and infects mechanically damaged corms during their storage in a bulb shed. Corms that are damaged at the end of the storing period are more susceptible to mildew. The corms develop yellowish white or light brown spots which grow in size during storage. Seriously infected corms do not form shoots and are easy to detect before they are planted. Slightly infected corms continue to rot in the ground; if they form new shoots, they remain small and dry up quickly, producing very small replacement corms which, like the neighboring plants, remain uninfected. The most important protection measure is careful harvesting and storing of corms in a well-ventilated shed.

Gray mold or **botrytis** is caused by *Botrytis croci* and *B. gladiolorum*. Seriously infected corms unusually have dark brown tunics, sometimes with small flat sclerotia on them. The upper part of the corm is covered with small round pinhead-size blackish brown spots. In a heavy infection, the shoots rot before emerging above ground, but in most cases the cataphylls become brown, the flower buds sometimes rot, and only green leaves appear above ground. Later on the leaves develop yellowish brown or brown spots of various sizes. In moist conditions dead flowers are covered with gray "wool." Small black sclerotia are formed on dead leaves. *Botrytis* is spread by infected corms and by spores. Because sclerotia have a short lifespan in the soil, crop rotation is essential to prevent the spreading of the disease (it affects other members of the family Iridaceae, too). Gray mold is more infectious in wet weather and it is very important to remove all faded flowers, especially on autumn- and winter-blooming species when the weather is dull and damp. All infected corms must be destroyed, others treated with fungicides. If the weather during the active growth period is rainy, it is essential to spray plants with a fungicide.

Fusarium rot is caused by *Fusarium oxysporum*. A closely related fungus causes pseudo-rust. In the case of fusarium rot, the infection begins in the basal plate and the corm becomes soft and yellowish brown. At harvesting time infected corms are easily recognized by a dark brown tunic that quite often is coated with stuck soil particles. When infected tissues harden and mummify, they usually are covered with white or pinkish mold. Such bulbs have loose-covering tunics and most of them harden (become stonelike) during storage. *Rhizoglyphus* mites live on the damaged corms and emit a specific scent. Sclerotia can remain in the soil for several years and mostly affect plantings which are overfertilized with nitrogen. Fusarium rot can quickly spread in the bulb shed if the air is humid and warm (above 21°C).

In the case of **pseudo-rust** the symptoms are closer to those of true rust. Large dark brown spots develop on the corm base or sides and penetrate as deep as the core of the corm. When planted, infected corms die or make weak shoots which remain compressed in shape (retaining vertical position) and prematurely wither. Unlike fusarium rot, pseudo-rust does not cause soil particles to adhere to the corm and mold is not formed, making it more difficult to find the infected corms.

To guard against both forms of *Fusarium* the grower should be cautious with fertilizer, carefully check the planting material, and destroy all suspicious corms. Crop rotation (six-field system) is essential.

Rust is caused by *Uromyces croci*. Symptoms on cataphylls and the bases of the green leaves are grayish to black-brown stripes that spread parallel to the veining. Later orange pustules of rust develop on large black spots. Spots are rarely seen on the aboveground parts of leaves, but they can develop on the corms, in the grooves where the covering tunics are attached. This disease can be controlled by hot water treatment—2.5 hours at 43.5°C—and by destroying all corms with spots and weak shoots.

Root rot can be caused by three agents. In the case of *Stromatinia gladioli*, roots develop light brown stripes and spots in early spring. Leaves turn yellow starting from the tips, and roots rot off completely. Sclerotia in the form of small black spots develop on cataphylls, and brownish black sunken spots form on the bottom of corms. The sclerotia stay in the soil for many years. Fungicide treatment and changing (or disinfecting) the soil can help.

The most serious root rots are caused by various *Pythium* species. The first symptoms may show up very soon after planting. Root tips become dark brown or blackish and easily break. Later the roots become light gray and transparent and their "skin" is easily to peel off. Heavily infested cataphylls develop ovoid or irregular spots with a darker edge. In spring when the weather is dry and warm the leaf tips grow yellow early in the season, and lead-gray spots appear on the leaves. The rot is spread by infected corms and remains in the soil for many years. Dutch growers use soil fungicides shortly before planting. Chemical treatment of harvested corms can help, too, but most of the specific fungicides against *Pythium* are now forbidden in the European Union. The only remedy remaining is to replant corms in clean soil.

A third cause of root rot is the root nematode *Pratylenchus penetrans*. Plant development is hindered, and plants start to wilt and turn yellow. On the roots one can see small, sharply marked brown spots which later become large and smudgy. Since the roots have died off, the replacement corms are very small but they are not infected. It is recommended to grow French marigolds (*Tagetes patula* or *T. erecta*) in the field one year before planting crocuses or to disinfect the soil with nematicides.

Gray bulb rot caused by *Rhizoctonia tuliparum* generally infects only leaves, while corms and roots remain healthy. The underground parts of cataphylls become partly or entirely brown and on them develop holes with fringed edges. The tips of the green leaves are light brown and usually adhered together. In a serious infestation, plants don't come up at all. In a minor infestation, plants recover when the soil temperature rises. When the disease is found for the first time, dig up the infected plants together with the soil around and underneath the corms and also the neighboring plants. Deep trench-digging of the soil can help.

Viruses

Crocuses are susceptible to viruses, but fortunately in most cases it is easy to spot the infected plants. Flowers of virus-infected crocuses will not open even in full sun, tepals look somewhat crumpled, and the tepals of dark-colored varieties have darker or lighter colored areas (Plate 12).

At one time I had a great collection of large-flowered Dutch crocuses (*Crocus* ×*cultorum* or *C. vernus*). When I looked more carefully at the leaves, I found that they had short yellow stripes parallel to the venation. I immediately thought of virus infection, but all the plants grew perfectly and flowered nicely from year to year. All my new imports from the Netherlands had the same sort of leaves. Then I collected seeds from these cultivars and raised my own seedlings. The first flowering brought me a big surprise: the flowers of my seedlings were much larger and they bloomed much more abundantly than I had ever seen previously on the Dutch varieties, and their leaves were without the yellow stripes. This confirmed my suspicion about the infection of Dutch stocks. I didn't separate my seedlings by color, but now in my garden I grow only my own hybrid seedlings of *C.* ×*cultorum*.

Many viruses do not damage the host plant seriously. The two organisms have learned to coexist. Such viruses are referred to as latent, or symptomless, viruses. They only become problematic when a plant is infected by a second virus. The addition of a second virus speeds up the destruction of the plant, which was previously coexisting with a single virus.

Iris gray-striped virus deforms crocus flowers, making them look compressed. Blue and purple flowers have dark stripes and spots. Three to four weeks after plants infected with this virus flower, the leaves develop light and dark green stripes, starting from the base. When heavily infected, the leaves have narrow variously colored stripes.

Mosaic can be caused by the cucumber mosaic virus and by the bean mosaic virus. In the case of cucumber mosaic, the outside of the tepals develops dark, very thin discontinuous stripes, the color whitens. With a serious infection, tepals can curl up. In the case of bean mosaic, the leaves develop light and dark green mosaic-like patterns once flowering is ended (Plate 13). Mosaic is primarily found in yellow-flowered crocuses, where no symptoms can be seen in the flowers themselves. It sometimes infects *Crocus sativus*.

Tobacco rattle virus is very infectious and it is spread by non-parasitical soil-inhabiting nematodes. The leaves develop yellowish green and/or unglazed white stripes and/or diamond-shaped spots; sometimes the leaf edges are serrated, and flowers can be striped (Plate 14). Often symptoms are not visible on all the shoots emerging from the corm, a factor that makes this virus especially dangerous, as gardeners may be prone to assume that something other than a virus is responsible for the floral discoloration since not all the shoots and leaves look infected. This virus infects a wide range of plants, especially bulbs and perennial weeds.

Arabis mosaic virus appears on the tepals as discolored or darker spots that widen into a fan shape toward the edge of the tepal. The tepal surface becomes uneven. This is the only symptom of the virus in white-colored varieties. Leaves remain green and healthy-looking. Like the tobacco rattle virus it is spread by non-parasitical soil-inhabiting nematodes.

The only way to eliminate these viruses is to destroy the infected plants immediately after spotting an infection and maintain the plantings free from weeds and aphids, which I have never seen on my crocuses but which are reported to be one of the vectors in the spreading of virus infections in the field and the bulb shed. Infected plants and corms must be burnt. Species can be revitalized (cleaned from infection) by sowing seed. Most viruses are not transferred through seeds, although sometimes two generations are needed. Problems develop when a virus causes plant sterility and the infected specimens don't set seed.

Non-parasitic Problems

Zigzag-leaves is an anomaly in which the tips of green leaves grow in a zigzag pattern. Flower tepals can be damaged as well, and sometimes flowers are malformed, paperlike, or dry. Sometimes only one shoot of a corm is affected, and all the other shoots are normal. In the 1980s I brought such plants to the virus research laboratory at the Latvian Agricultural University, but no virus infection was found. The cause of these symptoms remains unknown. It may be that the frozen soil surface doesn't allow the fast-developing shoot to squeeze through the hard cover, so the leaf tips in these circumstances grow irregularly.

Sun burn occurs when harvested corms are left out in the sun for too long, and the parts of the corm that are not covered by the tunic are damaged. Likewise, some species with thin tunics can be damaged even if the tunics are harvested intact. The color of the affected corms is grayish or brownish blue. Seriously affected corms dry up and die. The only way to protect corms is to move them into shade as soon as possible after lifting them from the ground.

Like other bulbs, crocuses are susceptible to **frost damage**. Dutch researchers have shown that crocus corms can tolerate temperatures as low as –2°C at the planting depth for several weeks. At colder temperatures the corms can die after a few days. Additional factors that influence hardiness include the soil structure, moisture level, planting time and depth, mulch, and snow covering. After late frosts crocus leaves become brown starting from the tips. Roots develop brownish streaks that can easily be misidentified as *Pythium* infection. The tissue at the corm base and in the core becomes white and dry and sometimes a cavity (called a "hollow heart") forms.

In Latvia the soil has several times been frozen to a depth of 1 meter, but crocuses didn't suffer. Usually crocuses are hardy here in winter, but when comparatively warm weather is followed by a hard frost, damage sometimes occurs. In the winter of 2005–06, the temperature reached –15°C one week in November, after which the temperature rose to +10°C, making the bulbs think that had spring had come. By January, many outdoor-growing crocuses were in full bloom. I have never had such a mass flowering of crocuses in January. But then the real winter set in with temperatures during all of February from –25 to –35°C. The first week was the coldest and completely without snow. When spring came that year, the crocus field was black. Only *Crocus heuffelianus*, *C. malyi*, and *C. veluchensis* survived those conditions.

Black frosts are most dangerous for crocuses, but a thin blanket of snow can protect plantings. In December 1978 the temperature in my garden at snow level dropped to –47°C but all crocuses survived because the soil was covered with 7 cm of very powdery snow. Even a few corms of *Crocus heuffelianus* that were left unplanted on top of the soil and covered only with snow survived winter and flowered the following spring.

PART II

Species of Crocuses

Botanical Characters

AROUND 100 *CROCUS* species are known, and new species are still being recognized. It is difficult to specify an exact number of species because various authors interpret differently the taxonomic status of many taxa, referring to them as separate species or only as subspecies or varieties.

Crocuses are widely distributed in the Old World starting in Morocco in North Africa and in Portugal in the westernmost part of Europe and reaching as far as eastern Kyrgyzstan and the Dzungarian Ala Tau in western China. Northward they reach southern Poland where *Crocus heuffelianus* subsp. *scepusiensis* occurs. The southernmost locations are in southern Iran, southern Jordan, and northern Libya. Most species fall within the Mediterranean region, but the largest number of species come from Turkey which can be considered the center of variability and distribution of crocuses.

To be able to correctly identify a crocus species, you have to know the botanical characters which are used to separate the species. The problem is that the plant is very small and several important characters lie underground. Therefore, you will have to lift a crocus plant at flowering time and dissect it, which means that the plant will perish.

The best way to learn the morphology of crocuses is to obtain several examples of the widespread and cheap commercial cultivars—one *Crocus vernus* (large Dutch) variety, one *C. chrysanthus* hybrid, and one representative of yellow *C. flavus*. All of these are available at supermarkets and are very cheap; thus your studies will not cost much money. Figure 2 shows the various parts of a crocus plant (left side of figure) as well as those of the autumn-blooming colchicum (right side of figure), which are often referred to as "crocuses."

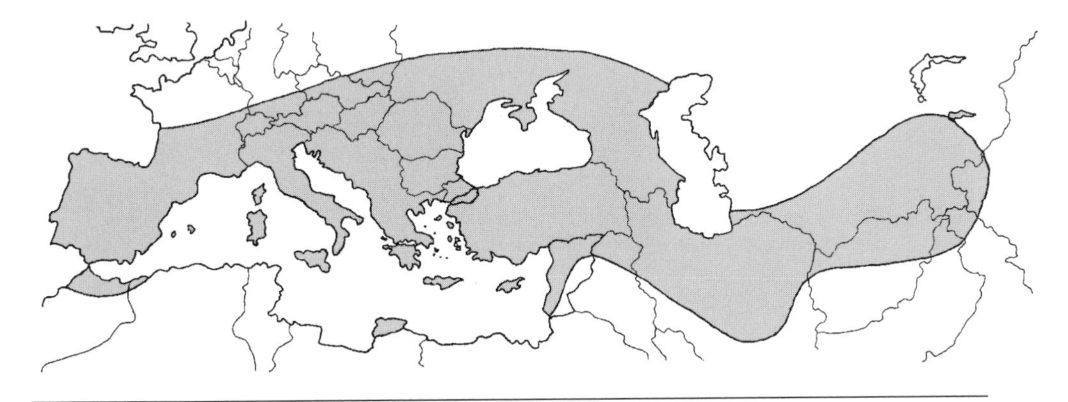

Figure 1. Distribution of crocuses in the wild.

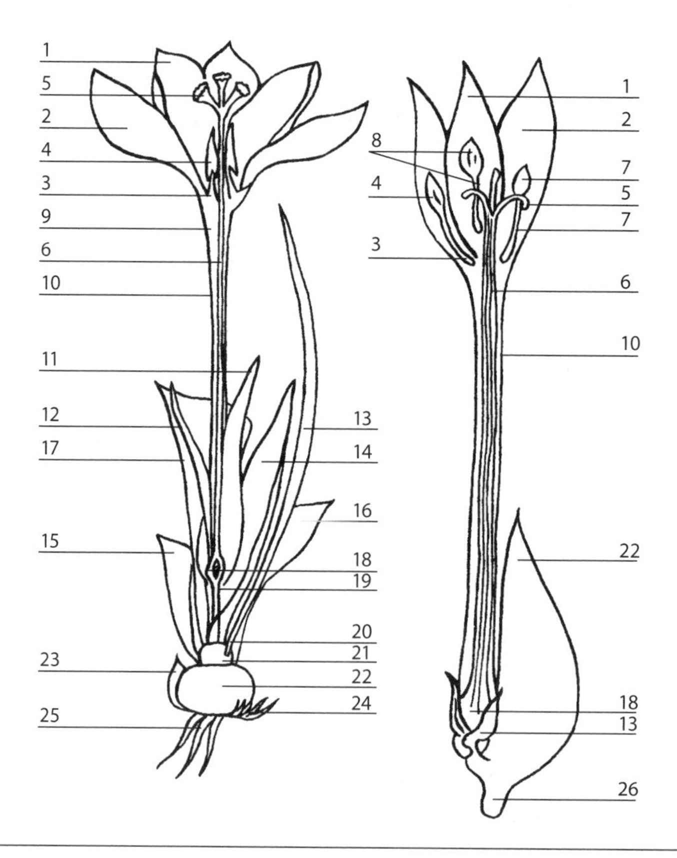

Figure 2. Plant and flower details of *Crocus* (left) and *Colchicum* (right). **1:** Petals. **2:** Sepals. **3:** Filaments. **4:** Anthers. **5:** Stigma. **6:** Style. **7:** Lower stamens. **8:** Higher stamens. **9:** Throat. **10:** Perianth tube. **11:** Bract (outer proper spathe). **12:** Bracteole (inner proper spathe). **13:** True (assimilating: leaves in *Crocus*, primordiums of true leaves in *Colchicum*). **14:** Prophyll or basal spathe. **15, 16, 17:** Sheathing leaves or cataphylls. **18:** Ovary. **19:** Scape. **20:** Leaf base. **21:** New corm. **22:** Old corm. **23:** Corm tunics. **24:** Basal tunic. **25:** Roots. **26:** Corm foot in *Colchicum*. Adapted from E. A. Bowles, 1952.

The crocus **corm**, which is the object that you usually buy, is a storage organ consisting of an almost uniform mass of tissues. Actually it is a much reduced one-year-old stem; on the top of this stem is a bud or shoot. At the end of vegetation, a new corm will form at the base of the bud or shoot. The old corm uses all its food reserves to produce leaves, flowers, and new corms before shriveling up and dying. Sometimes more than one bud is produced, in which case each of them forms a new corm. In some species at the base one or more tiny new corms, called cormlets, will form at the end of the growing season.

The corm is covered with dry tunics, which are the leaf bases from the previous season (Plate 15). A corm differs from a true bulb by being a solid body, while a bulb consists of scales in which food reserves are stored in the bases of the previous season's leaves and where the stem is present only as a small basal plate. You can easily see these scales when you vertically cut open a tulip or a daffodil bulb.

Tunics vary so significantly that they are important features in identifying species. When you carefully check the tunic of any *Crocus vernus* variety, you will see

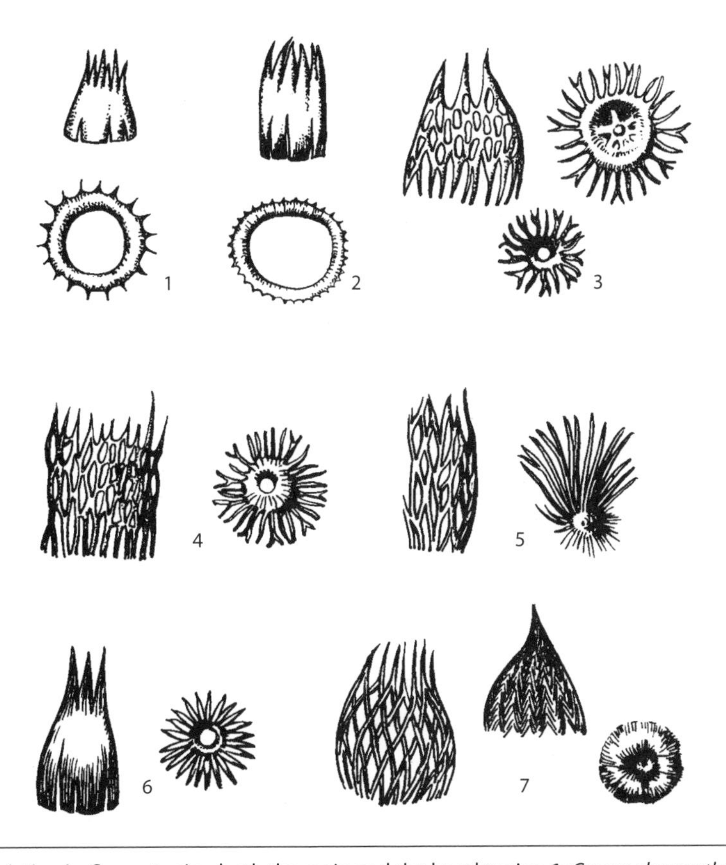

Figure 3. Variation in *Crocus* tunics, both the main and the basal tunics. **1:** *Crocus chrysanthus*. **2:** *Crocus melantherus*. **3:** *Crocus ancyrensis*. **4:** *Crocus etruscus*. **5:** *Crocus ligusticus*. **6:** *Crocus olivieri*. **7:** *Crocus fleischeri*. Adapted from G. Maw, 1886.

that it consists of fibers arranged in a fine net. In *C. angustifolius* the netting is very coarse and very prominent. *Crocus sieberi* and many other species have similar coarsely reticulated tunics.

The tunic fibers of *Crocus flavus* run parallel from the top of the corm to the bottom and are not netted. They are persistent and when you carefully remove them, you will see that they actually are bases of old leaves. In the wild they do not decay and each season new layers of dry leaf bases push the older layers up.

Crocus chrysanthus has a very special tunic (Figure 3). It is smooth without fibers and forms a round cap on top of the corm. Around the base of the corm are basal tunics, which in the case of *C. chrysanthus* are ringlike.

Crocus fleischeri has a very unusual tunic (Figure 3). Its fibers are distinctly plaited together. Once you see it, you will never confuse this species with any other. In some species fibers for the most part are parallel but at the tip they become finely reticulated, while in some others they are coriaceous (tough) but split at the base.

In some cases the position of a corm in the ground is of importance. Corms of most crocuses lie horizontally (upright), but some subspecies of *Crocus kotschyanus* can be identified by corms which lay vertically (Plate 16). The corms of *C. suworowianus* also lay vertically in the ground (Plate 17). The shape of crocus corms can also

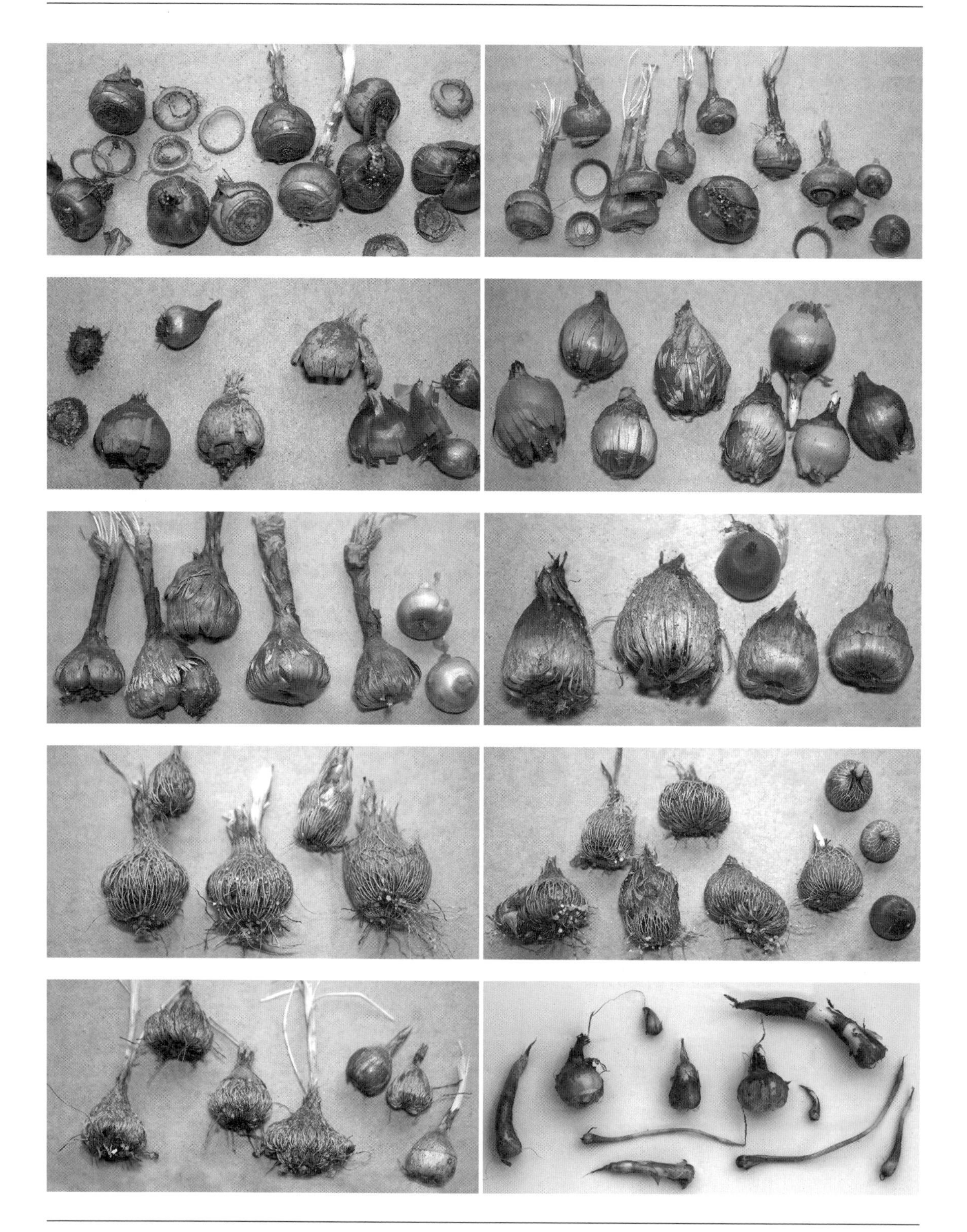

Figure 4. Variation in *Crocus* corms. Membranous tunics with basal rings (row 1), *C. chrysanthus* (left) and *C. danfordiae* (right). Membranous tunics without distinct basal rings (row 2), *C. wattiorum* (left) and *C. goulimyi* (right). Parallelly fibrous tunics (row 3), *C. flavus* subsp. *flavus* (left) and *C. aleppicus* (right). Reticulated tunics (row 4), *C. reticulatus* (left) and *C. ancyrensis* (right). Interwoven fibers (row 5, left), *C. fleischeri*. Stoloniferous corms (row 5, right), *Crocus nudiflorus*. Photos 1 to 9: M. Kammerlander; photo 10: J. Rukšāns.

vary—there are flattened or ovoid or subglobose corms; they can be round or somewhat irregular in shape.

At the bottom of the corm you can see a rounded part. That is the basal plate to which the old corm is attached. At flowering time two corms are present—the corm of the previous year, which becomes soft and wrinkled, and the new corm, which is still small, white, round, and firm. The old corm is easy to remove at the end of vegetation.

Roots develop in autumn around the basal plate of the corm. Usually they die off at the end of vegetation. In a few species, such as *Crocus veluchensis* and *C. scharojanii*, roots of the previous season coexist with the roots formed by the new corm. Usually roots are unbranched but in a few species (for example, *C. veluchensis* and *C. vernus*) they branch quite freely.

Young seedling corms and cormlets produced at the bottom of an old corm can form a contractile root which pulls the young cormlet down to a proper depth in the ground. This root is thick and very rich in starch. The contractile root transfers the starch to the young corm and shrivels, and at the same time pulls the young corm deeper. Very few crocus species form stolons which develop new cormlets at their tips in some distance from the mother corm.

The new stem and corm are enclosed in **sheathing leaves** or **cataphylls** (a botanical term for the early leaf-forms produced in the lower part of a shoot, such as bud-scales, or scales on underground stems). They can be three to five in number, papery, and tubular, whitish but often tinged greenish at the apex. The inner cataphylls are longer than the outer ones. Their bases form the covering tunics of the new corm.

The cataphylls enclose the **true** (that is, the green) **leaves**. They can be present at the time of flowering or absent, in which case they develop some time after flowering. This feature is important in identification of the autumn-blooming crocuses. Species which come from colder areas and high elevations usually bloom and overwinter without leaves. Examples are *Crocus speciosus* and *C. pulchellus*. They bloom in autumn but the leaves usually are formed only with the returning warm weather in spring. This kind of leaf formation is called hysteranthous. The autumn-bloomers which form leaves in autumn synchronously with flowers or at least at the end of flowering are called synanthous and come from warmer regions with milder climate and lowlands. Examples of these are *C. goulimyi*, *C. pallasii*, *C. melantherus*, and many others. They overwinter with green leaves and thus in my area can only be grown under cover. Spring-flowering crocuses develop leaves before or simultaneously with flowers.

The inner tunics of a new bulb are formed from the bases of green leaves. The majority of crocus species have leaves with a pale or white stripe. In a few species, such as *Crocus scardicus* and *C. pelistericus*, this stripe is absent. The shape, color, and number of leaves are very important taxonomic features. The basal portion of the leaf is colorless and strap-shaped, but higher up the leaf widens and both leaf margins roll back. A cross section of the leaf consists of two widened arms between which on the upper side is the aforementioned pale stripe but on the underside is a prominent keel. Roughly the leaf design seems to have something of a T shape.

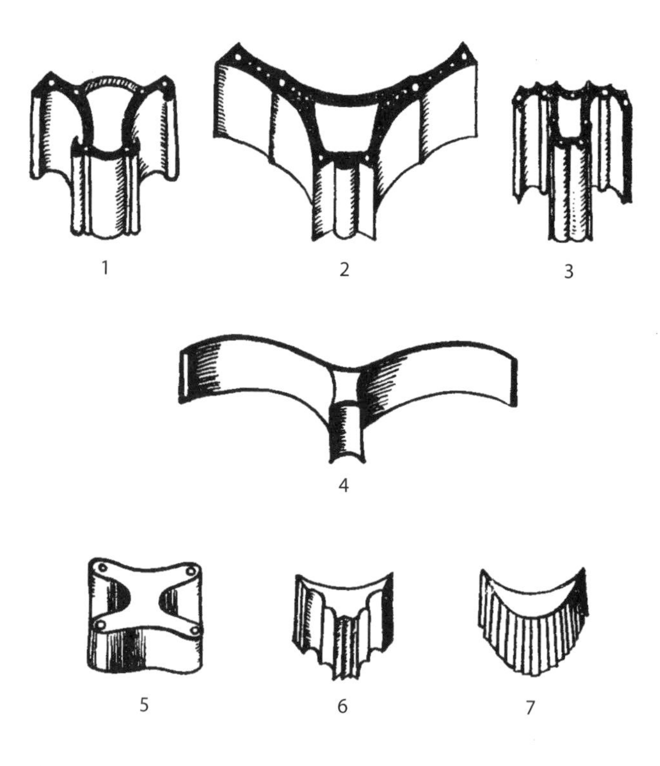

Figure 5. *Crocus* leaves, in cross section. **1:** *Crocus sativus*. **2:** *Crocus vernus*. **3:** *Crocus nudiflorus*. **4:** *Crocus banaticus*. **5:** *Crocus vallicola*. **6:** *Crocus nevadensis*. **7:** *Crocus carpetanus*. Adapted from G. Maw, 1886.

In some species there is a fringe of fine hairs on the edges of the keel and blades. Such leaves, called ciliated, are seen in *Crocus flavus*. Leaves without hairs, such as those of *C. vernus*, are described as glabrous.

The leaves of some species differ from the common mold. *Crocus carpetanus* (Figure 5) and *C. nevadensis* (Figure 5) have leaves that are semicylindrical in cross section with several small grooves on the underside. In leaves of *C. vallicola* (Figure 5) the keel is almost as wide as the arms and in cross section the leaves look square. Some species have grooves or ribs on the underside of the wings, and their absence or presence and the number have a certain value in the determination of species.

The **scape** appears above ground only at the fruiting stage when it carries the seed pod on its top. It is unbranched but can have several scapes inside each aerial shoot so that several flowers may be formed. In some species the seed pods remain underground and are not seen unless the corm is dug up.

The **prophyll** or **basal spathe** is a papery leaflike structure attached to the base of a scape. It is formed adjacent to the newly developing corm and it wraps the basal part of the scape. If there are several scapes they all are wrapped inside one prophyll. Species with a basal spathe (for example, *Crocus vernus*) are assigned to section *Involucrati* or, according to the current taxonomical rules, section *Crocus*. Species without a prophyll (for example, *C. flavus* and *C. angustifolius*) form section *Nudiscapus*.

Other papery structures form at the juncture where the ovary is connected to the scape. In some species only one papery leaf, called a **bract**, is attached; in others another smaller one, called a **bracteole**, is present. If several scapes are formed, each has its own bract and bracteole (if present). The bract or proper spathe is generally clearly visible above ground as is the bracteole when present. The absence or presence of bracteole is an important practical feature in the determination of species. Species without bracteole are called monophyllous; those with bracteole usually are called diphyllous.

At the top of the scape is the **ovary**. It is trilocular with two rows of seeds in each loculus. For the most part the ovary stays underground, emerging above the surface or rising higher only when the seeds are nearly ripe. Atop the ovary sits the **perianth** formed by a long **perianth tube** which instead of the stem raises the colored flower parts out of the soil. The perianth tube is formed by coalesced basal parts of the **perianth segments** (tepals). The flower is composed of a perianth tube and six perianth segments arranged in two whorls of three segments each—an outer whorl (sepals) and an inner whorl (petals). The variation in color, shape, and size is enormous. Even within one species there are a great number of color variants and hence this feature is of lesser taxonomic value.

As a rule the outer segments are longer and more pointed than the inner ones. In many cases the outer segments are slightly different in coloration as well and the exterior of outer segments can be conspicuously striped or suffused with brown or purple. Duller or darker outer coloration makes an unopened bud less conspicuous, which thus provides some protection from the insatiable herbivorous animals on mountain slopes in their homeland.

Crocus flowers usually are photoactive. At night and in dull cloudy weather the flowers remain closed because the natural pollinators are not active then and it is better to stay unseen. The flowers of *Crocus alatavicus* are almost invisible when closed, but with the first rays of sunlight, the slopes, which a few moments ago looked empty, suddenly look like they are covered in white snow when the crocuses start to open. As the sun rises and temperatures increase, the flowers quickly open to show the bright inner surface of perianth segments which can attract various pollinators. Only flowers of *C. tournefortii*, *C. cartwrightianus*, and *C. moabiticus* remain open at night. It is assumed that they are pollinated by a night or late-evening flying moth.

The point where the perianth segments merge and form a tube is called the **throat**. The inner surface of the throat can be hairy (called pubescent) or nude (glabrous). This feature, too, helps in the identification of crocus species. The throat can be white or yellow to orange in most cases, but some species have forms with a brown, blue, or even deep purple throat. The throat of some species is the same color as the inside of the perianth; in others the colored zone takes the form of a ring or blotches.

Stamens consist of **filaments** and **anthers**. The anthers usually are longer than the filaments (about twice as long). They open extrorsely (with their backs to the flower center) and this is characteristic of all crocus species included in subgenus *Crocus*. Only in *Crocus banaticus* do the anthers open introrsely (with their backs to the flower outside), and for this reason Brian Mathew placed this species in a monotypic subgenus *Crociris*. The size of filaments, their pubescence and color are used in identifying species. In most cases filaments are more or less papillose but in a few species they are hairy. If the throat is yellow, the filaments usually are yellow, too, but in *Crocus cyprius* they are deep orange or even reddish. When yellow is absent in the throat, the filaments can be white or creamy or even lilac-tinted.

Anthers are mostly yellow or orange; in several species they are white though sometimes they are black or have gray or blackish connectives. The color of the anthers must be ascertained before their dehiscence as pollen grains in most cases are yellow regardless of the anther color and after opening they hide the original color of the anthers. In many forms of *Crocus chrysanthus* the yellow anthers show black bottom tips.

The degree of branching of the **style** is a very important and easily visible taxonomic feature. J. Gilbert Baker (1873) separated crocuses into three sections by the degree of the style division—*Holostigma* with an entire stigma, *Odontostigma* with toothed or slightly divided stigmatic branches, and *Schizostigma* with a deeply divided

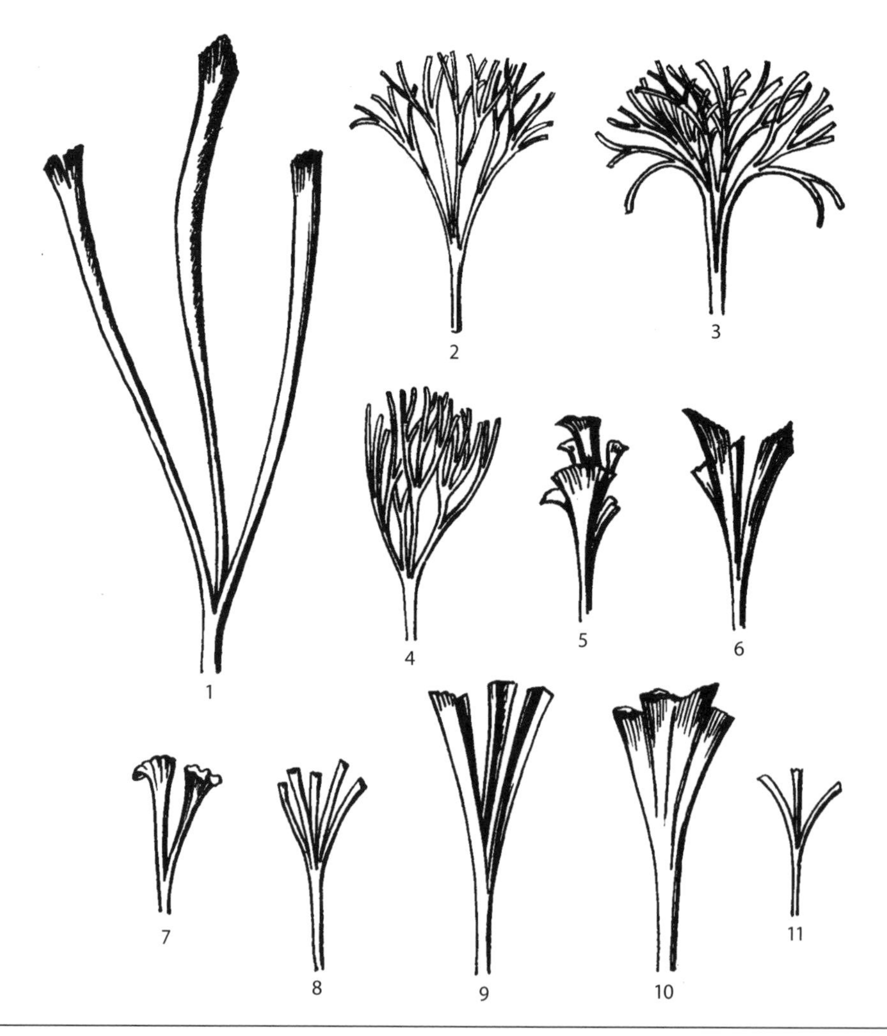

Figure 6. Variation in *Crocus* stigmas. **1:** *Crocus sativus*. **2:** *Crocus banaticus*. **3:** *Crocus ligusticus*. **4:** *Crocus laevigatus*. **5:** *Crocus nevadensis*. **6:** *Crocus flavus*. **7:** *Crocus carpetanus*. **8:** *Crocus olivieri* subsp. *olivieri*. **9:** *Crocus biflorus*. **10:** *Crocus vernus*. **11:** *Crocus caspius*. Adapted from G. Maw, 1886.

and branched stigma. The simplest mode of division is that of three branches, but there are species with as many as 30 to 40 slender branches.

The color of style branches is used for identification of species or subspecies in some groups of crocuses. The style branches can be tinted from white through yellow to red, sometimes lilac, and in a few forms the stigmas are black. In most species the stigmatic branches overtop the anthers. The place where the style divides and its relation to the anthers are used in identification of some species. Sometimes the stigma sticks out when the flower is still in bud. The stigma is fertile from the first day of the flower's life.

Flowers of many crocuses are very fragrant at flowering time. When my collection is in full bloom, my greenhouse is flooded with the nice honey aroma. There is plenty of nectar in the throat of the flower to attract the attention of various pollinators.

After pollination the fertilized ovary develops into a **seed capsule**. The flower stalk starts to elongate and brings the young capsule to the soil surface where it stays among leaves receiving all the available light and warmth. In very few species the capsules remain underground, lying just under the soil surface even after the seeds ripen, and in a few species the capsules are pushed up to 20 to 25 cm above ground.

Seeds ripen at the end of vegetation. The valves of the capsule spread widely or even reflex, and the seeds fall out and are dispersed mainly by ants. Seeds are variable in size and color. Seeds of most species have a fleshy attachment, called a caruncula, that is rich in albumins and attracts the ants. The most seeds I have ever found in one capsule of an outdoor-growing crocus without an artificial pollination was 37 (in *Crocus korolkowii*). The number of seeds depends on the species and the weather conditions at the time of pollination. Sometimes there is only one or no seed in the capsule.

Chromosome studies have shown very great variation even within one species. Numbers of chromosomes found in crocus are $2n = 6, 8, 10, 12, 14, 16, 18, 20, 22, 24, 26, 28, 30, 34, 44, 48, 64$, and 70.

⌘ Classification

WE CAN LOOK at plant classification from two viewpoints. One is strictly scientific and is the domain of scientists and professional botanists. The other is a practical viewpoint in which gardeners put the findings of scientists to good account in a way that is accessible and useful for their purposes.

The foundation of the now-accepted classification of *Crocus* was laid by George Maw (1886), and the major divisions adopted by Maw may be convenient enough as aids in identification, but the system is far from a more natural classification in which more aggregations of closely related species have to be included. The most recent botanical classification of the genus was made by Brian Mathew after years of research in the field, garden, and laboratory. It was published in his marvelous monograph *The Crocus* (1982). Mathew divides the genus *Crocus* into two subgenera—subgenus *Crocus* and the monotypic subgenus *Crociris* with one species, namely, *Crocus banaticus*. Mathew further split subgenus *Crocus* into two sections according to the presence or absence of the prophyll (basal spathe). Species with a basal spathe he places in section *Crocus*, those without a basal spathe in section *Nudiscapus*. Then he groups the *Crocus* species into series—six (at present one more series is added) in section *Crocus* and nine in section *Nudiscapus*. The series comprise closely related species.

Mathew's botanical classification of the genus *Crocus* is now generally accepted and used by both scientists and gardeners. His monograph includes 80 *Crocus* species and several subspecies. Since the publication of his monograph, many new species and subspecies have been described.

Today botanists use modern technologies more extensively in their research. Chromosome studies and biochemical studies are done, and in many cases they bring about revolutionary changes in the understanding of the evolution of plants and their relationships. In 2008 the journal *Taxon* published the results of studies on nucleotide sequences in the genus *Crocus* carried out by Gitte Petersen, Ole Seberg, Sarah Thorsøe, Tina Jørgensen, and Brian Mathew. These studies sometimes conflict with Mathew's original classification of the genus. Among the greatest surprises in the phylogenetic tree of the genus *Crocus* compiled by the researchers are the findings that morphologically very different *C. banaticus* is closer to *C. malyi*, and *C. caspius* is very close to the Central Asian species although the corms look very diverse. Many inconsistencies were pointed out between the present classification and the new phylogenetic hypothesis, so a further reclassification would be welcome after more studies are conducted.

Mathew's monograph included an excellent key for the identification of *Crocus* species. With the discovery of new species, this key now needs to be expanded to include them. In most cases this key is very practical. The biggest problem is that sometimes it is necessary to know the exact origin of your plants in order to make the correct identification. For botanists working mostly with herbarium sheets or plants whose place of collection is precisely known, this doesn't create any problems. Unfortunately, gardeners rarely know such details. They obtain plants from nurseries, through plant exchanges, or from seed received in various ways. Such materials are not always correctly identified, in which case determining the name of the plant becomes problematic.

I once received a beautiful crocus from a Czech grower who reported it as a species collected in eastern Turkey. This crocus flowered nicely, and since I liked it very much, I tried, for two seasons, to identify it using the keys in *Flora of Turkey*. Unfortunately, no species in that key matched my plants. In the third season I used Mathew's monograph and immediately identified my plants as *Crocus nevadensis* from Spain! The name of Turkey on the label of the plant I received made me blind to the shape of the plant's leaves.

Another problem in identifying crocuses is the wide variability of color in some species. Some species have a very uniform color, but in most of them you can find different color forms, especially albinos. Any key of the genus *Crocus* can fail when trying to identify such forms. Yet another problem is that some plant features are hidden underground. The shape of a corm tunic can be seen at the time of planting, but the presence or absence of a basal spathe is only seen when the plant is dissected, so it is a much less practical feature for including in a key for gardeners.

One more problem is that plants are often incorrectly named when one receives them. Few genera have species with as many synonyms as does *Crocus*. One of the most well-known crocuses, *C. flavus*, was given more than 10 different specific epithets over the centuries it has been known, while the name *C. vernus* has been applied to more than 10 different species. This situation could have caused total chaos, but fortunately the *International Code of Botanical Nomenclature* now regulates the rules of naming a species. The basic principle is that priority is given to the name that was officially published first. At the time of writing this book the World Checklist of crocus names compiled by the Royal Botanic Gardens, Kew, listed 696 crocus names, of which only 142 are accepted at species, subspecies, or varietal level. Unfortunately, many popular plants in commerce still bear their old synonymous names. For example, in commercial catalogs, *C. angustifolius* is often offered under its rejected name of *C. susianus*. Garden cultivars and hybrids have a similar code to regulate the rules of their naming, the *International Code of Nomenclature of Cultivated Plants*.

As the results of molecular studies based on chromosomes and DNA sequences are incorporated in plant taxonomy, many well-established views on relationships in the plant world will need to change. However, gardeners need to refer to more discernable and tangible features when checking a plant name. Sometimes it is difficult to find the simplest differences between two species, and preparing good keys for plant identification in the field and in the garden is not easy. I still grow hundreds of samples under collector's numbers only, waiting for identification. Identification is especially difficult in cases where you have only one plant or clone, which can be a very special form. Good keys, a sharp eye, and a lot of time are needed here.

Gardeners usually use their own plant grouping systems including both botanical species and cultivars which are quite often obtained from crosses between different species. Those kinds of grouping systems have been worked out for many popular garden plants such as daffodils, lilies, tulips, and many others, though from time to time these groups are changed. Crocuses are no exception to this rule. The third edition of *Classified List and International Register of Hyacinths and Other Bulbous and Tuberous Rooted Plants* (Royal Bulb Growers Association 1975) divided *Crocus* cultivars into 16 groups. In recent editions these groupings have been omitted.

The principal feature by which gardeners group crocuses is flowering time. All crocuses can basically be divided into autumn- and spring-blooming species. This feature is more obvious in Latvia, for example, where real winters mark a sharp border between the two groups. In Great Britain and other countries where crocuses start blooming in early autumn and continue through winter until late spring, it is not that easy to draw a definite line. In such climates, the beginning of a new calendar year can be used to separate spring-bloomers from autumn-bloomers. Crocuses which start blooming in autumn before 31 December can be called autumnal bloomers, those that start later are spring-bloomers.

Unfortunately, such distinctions cannot always be made with ease. Autumnal species such as *Crocus melantherus*, *C. cambessedesii*, and *C. laevigatus* have forms which bloom only in spring. The Central Asian species which typically bloom in early spring, in some seasons can begin to flower in late autumn. The earliest record of blooming of *C. michelsonii* in my collection is from 15 November; *C. korolkowii* has started to flower at the end of December.

Some nurseries and bulb sellers divide the spring-blooming crocuses into three groups: the large Dutch crocuses (raised from *Crocus vernus* and including in this group the hybrid of *C. flavus* known as *C.* 'Yellow Mammoth'); the so-called *C. chrysanthus* cultivars, which in most cases are selections from and hybrids between different forms of *C. chrysanthus* and *C. biflorus*; and a third group including all other spring-blooming species. When I had these cultivars I followed exactly this grouping of crocuses in my collection.

In this book, I have arranged crocuses in practical groups of use to both amateur and professional gardeners. I have tried to combine the principles of E. A. Bowles with the system devised by Brian Mathew, adding my own experience and including the recently discovered species not listed in Mathew's keys. The basic division is according to flowering season. Therefore I was forced to split in different chapters some botanically close species, particularly in the *Crocus biflorus* group. *Crocus sieberi* has been split into two species. Likewise the *C. vernus* complex has been split up into three species (and possibly a fourth is needed).

The greatest problem I faced in writing this book is that there are some species which I have never grown nor seen in a living state. However, for the completeness of this work I included such species in both the keys and the descriptions, using the descriptions given by Brian Mathew in *The Crocus* and several publications of Helmut Kerndorff and Erich Pasche. Habitat characteristics and elevations at which crocuses grow in the wild are mostly given following Mathew's monograph with a few additions from my own observations.

✾ Autumn-Blooming Crocuses

In this book I have divided all autumn-blooming crocuses into two artificial groups taking into account only one feature—whether they are leafless during flowering or whether the leaves appear before the flowering ends. In cases where the leaves start forming during flowering, it was not easy to decide into which group such crocuses belong. Brian Mathew et al. (2009) write:

> The autumn flowering habit in *Crocus* is probably not such a fundamental difference and is most likely an indication of not-so-distant past climate change events, particularly in the case of those species which produce leaves at flowering time. .... Some species having this characteristic are flexible in their flowering time from late autumn to early spring.

This method of dividing the autumn-blooming crocuses has a very practical application for growers in regions with harsh climates like that of Latvia. The species that bloom without leaves can be grown more or less successfully outside, but those that form leaves before flowers can be grown only under cover. That does not mean that all leafless autumn-blooming crocuses will grow outside in harsh climates. Many of them are not hardy; several produce leaves in autumn but well after the blooming has started. Such plants generally are not suitable for outside gardens in harsh climates.

Leafless Autumnal Crocuses

The growing of any crocus starts with the planting of corms. Thus, the first thing gardeners see is not a flower but a corm. For this reason the identification key for this group of species is based on the shape of the tunic covering the corm. The one exception is *Crocus banaticus*. Its flower is so unique that it is impossible to misidentify this species once you see it.

1. Inner segments much shorter than outer segments .... *C. banaticus*
1. Inner and outer segments approximately the same length
    2. Corm tunics with obvious rings at base
        3. Flowers pure white .... usually white-flowered forms of *C. speciosus*, *C. pulchellus*, or *C. kotschyanus*
        3. Flowers in type forms lilac or purplish

4. Anthers white, throat with a deep yellow zone, filaments densely pubescent . . . . *C. pulchellus*
4. Anthers yellow, filaments glabrous or minutely pubescent . . . . *C. speciosus*
   5. Leaves formed in autumn but after the flowering . . . . subsp. *archibaldii*
   5. Leaves appearing only in spring
      6. Throat of perianth deep yellow
         7. Style branches overtop anthers . . . . subsp. *xantholaimos*
         7. Style branches below or at same level as anthers . . . . . . subsp. *archibaldii*
      6. Throat of perianth white or faintly yellow
         8. Corm tunic tough, style many-branched, usually equaling or exceeding anthers . . . . subsp. *speciosus*
         8. Corm tunic thinly papery, style divided into 6–8 branches, usually overtopped by anthers . . . . subsp. *ilgazensis*
4. Anthers black
   9. Yellow zone in throat of flower large, ranging from deep yellow to orange, flower tube with purple stripes or purple, style not exceeding anthers, flowers stay more or less funnel-shaped in sun . . . . *C. nerimaniae*
   9. Yellow zone in throat of flower small, flower tube white, style branches much exceeding anthers, leaf tips appear at end of blooming, flowers widely open in sun . . . . *C. wattiorum*

2. Corm tunics thinly membranous, weakly reticulated at the top
   10. Flowers yellow or yellowish . . . . *C. scharojanii*
      11. Perianth segments acuminate, corms up to 15 mm in diameter, not stoloniferous . . . . subsp. *scharojanii*
      11. Perianth segments acute to rounded, corms 5–10 mm in diameter, can be stoloniferous . . . . subsp. *lazicus*
   10. Flowers white, whitish, or lilac
      12. Flowers without yellow in throat
         13. Style branches white
            14. Style much exceeding anthers, flower segments pointed . . .*C. karduchorum*
            14. Style equaling anthers, flower segments rounded . . . . *C. kotschyanus* var. *leucopharynx*
         13. Style branches yellow or orange
            15. Style branches much exceeding anthers . . . . *C. autranii*
            15. Style branches equaling anthers . . . . *C. gilanicus*
      12. Flowers in throat yellow or with yellow spots or rings
         16. Flowers with long acuminate tips . . . .*C. vallicola*
         16. Segments obtuse or rounded
            17. Corm lying upright . . . . *C. kotschyanus* subsp. *kotschyanus*
            17. Corm lying on its side

18. Flowers lilac
    19. Throat glabrous, perianth segments obovate or oblanceolate .... *C. kotschyanus* subsp. *cappadocicus*
    19. Throat pubescent, perianth segments obtrullate .... *C. kotschyanus* subsp. *hakkariensis*
18. Flowers white or rarely pale lilac, throat glabrous, perianth segments lanceolate .... *C. suworowianus*

2. Corm tunics membranous with parallel fibers, corms stoloniferous .... *C. nudiflorus*
2. Corm tunics fibrous, splitting in fine fibers parallel at base and weakly reticulated at apex
   20. Style trifid .... *C. moabiticus*
   20. Style multifid, divided into 6 or more slender branches .... *C. hermoneus*
       21. Style clearly exceeding tips of anthers .... subsp. *hermoneus*
       21. Style not exceeding tips of anthers .... subsp. *palaestinus*
2. Corm tunics distinctly coarsely reticulated
   22. Style trifid .... *C. robertianus*
   22. Style many-branched
       23. Bract conspicuously green, bracteole absent .... *C. ligusticus*
       23. Bract white, bracteole present but may be hidden .... *C. cancellatus*
           24. Anthers white, throat deep yellow or orange .... subsp. *pamphylicus*
           24. Anthers yellow, throat yellow or almost white
               25. Bracteole visible, not hidden within the bract
                   26. Style equaling or shorter than anthers, flower segments up to 3 cm long .... subsp. *lycius*
                   26. Style much exceeding anthers, flower segments 3–5.5 cm long .... subsp. *mazziaricus*
               25. Bracteole hidden within bract
                   27. Corm with long bristly fibrous neck, tunic with coarsely netted fibers ....subsp. *damascenus*
                   27. Corm with very short bristly fibrous neck, tunic with more finely netted fibers .... subsp. *cancellatus*

I would like to start with the most unusual *Crocus* species not only in this group but also in the entire genus, namely, *C. banaticus*.

Crocus banaticus Gay

PLATES 18 & 19

Crocus banaticus ($2n = 26$) is a plant from meadows and deciduous woods and thickets growing from 130 to 700 meters. Its flowers are so distinct that it was once even separated into its own genus as *Crociris iridiflora* Schur. From this name you can picture that its flowers somewhat resemble those of an iris, although this isn't always exactly the case, with the inner flower segments being only half the length of the outer ones. On warm days when the sun is shining the outer tepals sometimes open widely, somewhat resembling the falls of irises, while the inner petals remain upright, looking like the standards of an iris bloom.

But this isn't the only peculiar feature of this species. In its typical lilac-purple form, the many-branched stigma usually is light purple. I don't know of any other crocus with such a color of the stigma. Its anthers open introrsely (facing the flower's center), while in all the other species they open extrorsely (facing the tepals).

All these features spurred Brian Mathew to separate this crocus into a special monotypic subgenus—subgen. *Crociris*. It was quite a surprise when studies on the molecular level showed that phylogenetically *Crocus banaticus* was somewhat close to a typical *Crocus* species—spring-blooming *C. malyi*—and that means that its separation from other species was not supported.

Over the centuries *Crocus banaticus* had been named several times and each time with a different epithet. The most appropriate name would have been *C. iridiflorus*, which perfectly describes its flower, but the priority goes to the name given by J. Gay—*C. banaticus*, which refers to its distribution in the province of Banat in Romania.

Geographically this species is distributed in Romania extending as far as the northeastern former Yugoslavia and entering the eastern Carpathian Mountains in northwestern Ukraine where it is very rare. I have not seen it during my travels there. Its rarity in the Ukraine is supported by its listing in the Red Data Book of the former USSR and of Ukraine.

Several selections have been made. E. A. Bowles mentioned a form with lilac flowers and a white stigma. Purple forms have been named. Even a bicolored form is known; it has deep purple outer segments and light lilac inner ones. Beautiful albino forms also exist. The best one is **'Snowdrift'** with large pure white flowers and equally white stigmatic branches (Plate 19). Unfortunately, it blooms so late that I've never gotten it to bloom outside, but in the greenhouse it is a wonderful pot plant and a show-winner. It is a good increaser and is easy to grow. Similar but with somewhat smaller flowers is **'First Snow'**, originally collected near Oradea in Romania and named in my nursery. It usually blooms even outside here. In the greenhouse its flowering ends early and when I start to worry that I've lost 'Snowdrift', its large flowers suddenly pop up in the pot.

It is not easy to grow *Crocus banaticus* outside in Latvia as it blooms quite late and isn't completely hardy here. I lost this beauty several times before I built my greenhouse. Now in pots it grows marvelously, and every year in late spring I take the pots outside as it becomes too hot for them in the greenhouse. Being a plant from woodlands, *C. banaticus* requires some shade and more moisture than most of the crocuses. Even after the leaves die down, it is essential to water the pots occasionally and not allow a complete drying out. This species prefers a somewhat more humus-rich soil mix and with a few exceptions is a rather moderate increaser vegetatively. If you do not want any special color form, the fastest way to increase its stock is by seeds.

Crocus speciosus Marshall von Bieberstein

Crocus speciosus, meaning "the showy crocus," is the easiest and most widely grown species in gardens. It has been divided into three subspecies, but a fourth one is described here.

Crocus speciosus subsp. *speciosus*

PLATES 20–22

Crocus speciosus subsp. *speciosus* ($2n$ = 8, 10, 12, 14, 18) is one of the few autumn crocuses that has survived in my garden for many years without any care. It is very widespread in the wild, growing all around the Black Sea from Crimea, stretching on through the Caucasus and northern Turkey into the province of Bolu, where forms with exceptionally large flowers are found around Lake Abant. Further west it is replaced by the somewhat similar but smaller *C. pulchellus*, but not long ago this species was found in a very disjunct location in the center of Greece, very far from the other known populations. Eastward *C. speciosus* enters northern Iran, where a very distinct form grows which produces leaves in the autumn after the end of flowering. In the wild it grows in woodland and on pastures or in alpine grassland from 800 to 2350 meters.

I came across the type subsp. *speciosus* some 30 years ago in Crimea in early September when my family and I were enjoying the nice early autumn weather on the stony beach of the Black Sea—the so-called velvet season when the water is still warm enough to swim in but the air is not as hot as in summer. On one occasion, I left my family and, with A. Yabrova of the Nikitsky Botanical Gardens in Yalta, went up to a yaila in search of crocuses. The season had been very dry and *Crocus speciosus* had only started to flower, but nevertheless I brought home a few corms. They turned out to belong to the earliest flowering form of this species. Normally *C. speciosus* starts to flower here in September; this one always flowered in the boxes way ahead of the replanting time, already in the first days of August. Actually it was the earliest of all the autumn crocuses, probably only *C. scharojanii* would have bloomed earlier.

The next time I encountered *Crocus speciosus* was in Georgia near Tbilisi, where the species was growing in great abundance under pine trees within the city limits near Lake Cherepash'ye. Many bulbs were growing around the lake, which is a very well known recreation area that can be reached by bus. The bus stops on a wide boulevard, near a ski lift. Within the first steps up the mountain, my companions and I noticed the first crocus leaves. They belonged to *C. speciosus*. There I collected only a few corms, but they turned out to be an amazingly nice form that I later named **'Lakeside Beauty'**, after the place where it was collected. It blooms particularly abundantly with very large exceptionally light silvery blue slightly deeper blue-veined flowers. Every visitor to my nursery points to this form as it is the best among my *C. speciosus* cultivars. I can only imagine how beautiful the slopes around Lake Cherepash'ye were in autumn when the crocuses were in full bloom. When I returned to this site 25 years later, in spring of 2007, big changes had occurred. The lakeside had low-trimmed grass, asphalted roads, large stands, and swimming lanes set up in the lake. Not a single crocus leaf could be found there any more, only a few weak *Muscari armeniacum* plants somewhere on the rocks.

Not long ago I returned to Crimea, this time in midautumn (October), with the intention of taking pictures of wild *Crocus speciosus* for this book. I went to the Tschatir-Dag yaila and saw plenty of plants. As is typical for this species, the plants were growing only in shaded spots—near karst depressions covered by trees or in small islands of trees usually in dense grass. The weather was not favorable for picture-taking; very low clouds were spread across the sky and from time to time it was drizzling. Only on the last morning before leaving the yaila did bright sun greet us and I could take some pictures. The plants were nowhere as abundant here as they were near Lake Cherepash'ye in Georgia or, as I discovered later, near Lake Abant in northern Turkey, where in spring the crocus leaves covered the ground like grass, under very shady hazelnut (*Corylus*) shrubs. During my next visit to the Tschatir-Dag yaila the following spring, I found that at lower elevations in a beech forest in somewhat more open sites the leaves of *C. speciosus* covered the ground like green grass.

Flowers of *Crocus speciosus* are very variable in color, but mostly they are shades of lilac and purple. I have never found white-blooming plants in the wild and my Armenian friend Zhirair Basmajyan, who specially examined many wild populations at my request, didn't find any either. He wrote to me (pers. comm.): "In Pambak area there are some variations of *C. speciosus* with very pale flowers, and even near white forms, which also have yellowish throat and creamy anthers but no one pure white comparable with Dutch cv. 'Albus'."

The very popular **'Albus'** is one of only two Dutch-raised cultivars of *Crocus speciosus* that I grow at present. The other is the deepest purple form of this species ever seen by me—**'Oxonian'**. It can be easily distinguished from other cultivars by its deep purple flower tube. I grow a few cultivars raised by Zhirair Basmajyan, such as **'Cloudy Sky'** (Plate 21) and **'Gornus'**, and several others raised by Leonid Bondarenko, namely, **'Lithuanian Autumn'** and **'Blue Web'** (Plate 22). The latter is quite small but most beautifully veined with deep purple lines on a light background. Most of these selections are good increasers in the garden as they form many pea-size cormlets at the base of the parent corm.

Crocus speciosus* subsp. *ilgazensis B. Mathew

PLATE 23

From *Crocus speciosus* Brian Mathew separated two new subspecies. One originally collected on Ilgaz-Dag in Turkey, he named *Crocus speciosus* subsp. *ilgazensis* ($2n = 6$, 8). It is quite easy to recognize by the stigmatic branches which are less divided and usually are overtopped by the anthers. The flowers of this subspecies are rather small compared with flowers of the other subspecies, and it is in fact a diminutive form of subsp. *speciosus*. It is a plant from alpine grassland and stony hillsides or clearings in *Abies* forest where it grows on limestone formations from 1600 to 1750 meters. Unfortunately, commercial growers do not always offer correctly named plants. In these cases I check the position of the anthers relative to the stigma.

Crocus speciosus subsp. *xantholaimos* B. Mathew

PLATE 24

The other subspecies that was named by Brian Mathew is easier to identify because it has a deep yellow throat. It is named *Crocus speciosus* subsp. *xantholaimos* ($2n = 10$). In the wild it grows in a rather limited area in the province of Sinop on mountain passes in clearings of *Abies* and *Rhododendron* woods at an elevation of around 1300 meters near the Black Sea coast. It is a very nice plant with thinly papery corm tunics and quite indistinct basal rings. You can find plants with a yellow throat in some populations of the type subsp. *speciosus*—I have seen such plants in Crimea and got several samples from Armenia—but doubtlessly, in the typical subsp. *xantholaimos*, the yellow color is deeper. Because color is quite relative—the difference between "deep yellow" and "only yellow" is subjective—the main feature is the shape of the corm tunic.

Crocus speciosus subsp. *archibaldii* Rukšāns, subsp. nov.

PLATES 25 & 26

> *Crocus speciosus* subsp. *archibaldii* Rukšāns, subsp. nov., subsp. *xantholaimos* (fauce intense lutea) et subsp. *ilgazensis* (antheras styli ramis aequus vel superans) similis sed foliis autumnalibus et flores colores ordinatione diversa (fasciatus non venosus) bone differt.
>
> Typus: Iran, Kuhha-ye Tales, between Nav and Khalkhal, steep mountain slopes just before pass, 2080 m. 2008-27-04, Rukšāns, WHIR-129 (GB, holo, ex culturae in Horto Jānis Rukšāns, flores 2008-09-15, folia et cormus 2008-10-20). Ic.: Crocuses: A Complete Guide to the Genus (Portland, OR, 2010), plates 25 and 26.

In spring 2008 I collected a few corms of *Crocus speciosus* at several spots in western Iran north of the city of Zanjan at elevations from 1800 to 2100 meters. The plants bloomed for me for the first time in the autumn of 2008 and immediately attracted my attention with a very unusual (for *C. speciosus*) flower color. The base color of the outer tepals on the outside was white and the tepals were covered with wide lilac stripes from the base to the tips. When the flowers opened they had another surprise in store—the veining on the inside of the tepals was something (not exactly) similar to the other subspecies of *C. speciosus* only they had a bright yellow base color as in subsp. *xantholaimos* and the position of the anthers relative to the stigma and the branching of the stigma were the same as in subsp. *ilgazensis*. The last surprise I got a week after flowering ended when the leaves started to grow; the plant entered winter with well-developed leaves whereas all the other *C. speciosus* samples didn't show even the tips of their leaves.

After an exchange of information with Brian Mathew we both came to the conclusion that it was a new, not-yet-described subspecies of *Crocus speciosus*. I decided to name it for the great explorers of bulbous and alpine flora who have introduced so many beautiful bulbs as well as crocuses through their seed distribution company not only from Iran but from very distant countries as far as Patagonia in South America—Jim and Jenny Archibald.

I collected the Iranian *Crocus speciosus sensu lato* in four localities; two samples were of the same appearance and characteristics as in subsp. *archibaldii*, but the other two looked like the traditional subsp. *speciosus*. The following spring I compared the leaves of all four gatherings. Leaves from samples regarded as subsp. *speciosus* were almost half as wide (2–3 mm) as those from samples of the new subsp. *archibaldii* (5–7 mm).

Crocus pulchellus Herbert

PLATE 27

Crocus pulchellus ($2n = 12$) replaces *C. speciosus* in northwestern Asiatic Turkey, and through the European part of Turkey enters the Balkans where it grows from northern Greece to southern Bulgaria and in southern former Yugoslavia. It is found in moist grassy places and in sparse pine or oak woods or scrub from sea level to 1800 meters where it experiences some summer rain. The name of this species means "small and beautiful," although it is not the smallest of crocuses; however, its flowers are smaller and more rounded than those of its close relative *C. speciosus*.

In their type forms, both species are easy to distinguish as *Crocus pulchellus* always has a deep yellow throat that is a darker shade than that of *C. speciosus* subsp. *xantholaimos*. Its anthers are white whereas those of *C. speciosus* usually are yellow and the filaments are densely pubescent. In cases where plant identification is doubtful—I have forms of *C. speciosus* collected wild in Armenia with a deep yellow throat and almost white anthers—I check the filaments. According to the new phylogenetic tree both species are allied to a number of species with a reticulated tunic (autumn- and spring-blooming) and to several species from the *C. biflorus* group.

Flowers of typical plants of *Crocus pulchellus* are of pale lavender shade, lack the cross veining of *C. speciosus*, and are marked only with a few slightly feathered darker bluish veins. Flowers are more globular and have shorter and stronger tubes. The corms are smaller and lack the grains at the base, and the flowering starts at least a fortnight later than on average is the case in *C. speciosus*. In the wild *C. pulchellus* is distributed on vernally damp meadows and doesn't like excessive drying out in summer, although it is a very tolerant garden plant and one of the few autumn-blooming crocuses that grows easily in Latvian gardens. Here it is somewhat less hardy than *C. speciosus*, and in severe winters *C. pulchellus* will suffer more seriously than *C. speciosus*.

The two species can hybridize quite easily in garden, especially when the later-blooming forms of *Crocus speciosus* grow alongside *C. pulchellus*. One such hybrid is **'Zephyr'** raised by Thomas Hoog and usually offered in catalogs as *C. pulchellus* 'Zephyr'. It has huge flowers of the size of *C. speciosus* and in fact looks more like a variety of it. It is fertile and when its seedlings started to flower with me, they clearly split into four forms, half of which looked like the typical 'Zephyr' but were more vigorous and floriferous than the mother plants. I think they were free from various viruses picked up during the long years of cultivation by commercial growers. I immediately destroyed my original stock of 'Zephyr' and replaced it with the seedling. Another hybrid which looks more like *C. speciosus* is named **'Big Boy'**; only because it has white anthers do I still keep it under *C. pulchellus*. Another cultivar, **'Late Love'** distributed by Leonid Bondarenko as *C. speciosus*, really is a hybrid with *C. pulchellus* too, but it looks closer to *C. pulchellus*.

The other half of the hybrid seedlings looked like *Crocus speciosus* and *C. pulchellus* approximately in equal proportions, but among the *C. pulchellus*-type seedlings were several with the purest white flowers. A white *C. pulchellus* was my long-wished-for dream plant because I had read in Bowles (1952) that "it is one of the most beautiful of early autumnal albino forms." And now I was growing exactly such a crocus.

Three autumn-blooming species from Brian Mathew's series *Biflori* have corm tunics similar to those of *Crocus speciosus* and *C. pulchellus*. One of them (*C. melantherus*) has well-developed leaves at flowering time and is therefore included in the next chapter. Of the remaining two species, *C. wattiorum* at first was described by Mathew as a subspecies of *C. biflorus* but was later raised by him to species status. The other species is the rather recently discovered *C. nerimaniae*. Both species come from southwestern Turkey.

Crocus wattiorum (B. Mathew) B. Mathew

PLATES 28 & 28

Crocus wattiorum ($2n = ?$) was named after Peter and Penny Watt who found the species in Turkey in 1986 on Tahtali-Dag. It grows in crevices of limestone rocks at the edges of pine woodland from 100 to 500 meters.

The flower segments open widely in the sun showing a beautiful bright lilac-blue shade with slightly darker veining. The color is lighter at the tips, but the throat is surrounded by a diffused darker lilac zone. The throat is small, glabrous, light yellow. The sepal outside is buff-toned with a longer dark purple central stripe and shorter ones on either side. The most impressive features are its blackish maroon anthers, which often have significant yellow basal lobes, and the long dark red stigmatic branches, which resemble those of the *Crocus sativus* group. The style splits well below the anthers, and at the tip the style branches are only slightly subdivided into two or three short lobes. In *Flora of Turkey* (Güner et al. 2000), the style branches are described as being much longer than the anthers, but in my samples they are more or less uniform and only rarely slightly exceeding the anthers. In any case the style branches are excellently displayed over the widely open perianth segments, giving the impression that they are very long. The flower tube in *C. wattiorum* is white. I haven't noticed any development of leaves at the start of flowering; the leaf tips appear above ground only at the end of blooming, therefore I am including this species in this chapter.

Crocus wattiorum is one of the most beautiful autumn-blooming species. It is extremely rare in cultivation; therefore little is known about its specific requirements. According to Peter and Penny Watt (2006) it grows in the wild on humus-rich, somewhat acidic soils on limestone. Some growers refer to it as difficult in culture. In my experience, it is a rather good grower in pots.

This crocus comes from lower elevations and requires a dry and hot summer rest. It readily sets seed and the first flowers can show up in the third autumn.

Crocus nerimaniae Yüzbaşioğlu

PLATE 30

Quite similar to *Crocus wattiorum* is another Turkish autumn-bloomer described only recently—*Crocus nerimaniae* ($2n = 14$). It was discovered in 2000 by professors from Istanbul University and named after one of them, Neriman Ozhatay, who was the first to find this beauty near the ruins of ancient Labranda from 450 to 700 meters.

Crocus nerimaniae is easy to distinguish from its cousin *C. wattiorum* by the deep violet striped perianth tube which in the former is white. Other distinctions exist. The outside of the lilac-blue sepals is without prominent stripes, but the most remarkable feature is the large flower throat which is yellow at the base and surrounded by a wide, deep orange-yellow zone. This coloring makes it even more impressive than *C. wattiorum* in the few flowering plants that I've seen until now. The tepals do not open as widely as those of *C. wattiorum* and the flowers are more funnel-shaped. The style branches are deep orange, somewhat subequal to the anthers, and frilled only at the tips but not subdivided. The leaves appear only in spring. Flower color reportedly varies in the wild from white to soft mauve. In autumn 2009 I observed very few white-flowering plants and a pair of specimens with blue-and-white striped flowers. Peter and Penny Watt write that in its *locus classicus C. nerimaniae* grows among very outstanding forms of *C. pallasii* with broad, almost pure blue perianth segments, the inside of which is so heavily veined that it creates almost an appearance of a purple throat. Growing alongside *C. nerimaniae* at Labranda is spring-blooming *C. biflorus* subsp. *caricus*, too.

I visited the *Crocus nerimaniae* locality east of Milosh in early summer 2008 where I saw a few plants over the ruins of Labranda, but later found it growing quite widely in the vicinity of ruins near low shrubs in clearings of a sparse pine forest. The place looked like a beautiful scenic park covered with low-grazed grass, huge rock boulders in fantastic forms, and with a beautiful stream running through it. Only a few places in the mountains have aroused such nice feelings in me as the locality of *C. nerimaniae*. I revisited the locality in autumn 2009, when *C. nerimaniae* was in full bloom and really was shocked seeing a very great percentage of virus-infected plants. There were groups with not one healthy-looking crocus. All suitable spots were occupied by thousands of bee hives. Bees usually are not reported as virus-distributing agents, but there were plenty of pine aphids. Maybe another greenfly is feeding on the crocus leaves and distributing infection.

My knowledge of *Crocus nerimaniae* cultivation is rather limited. The first plants bloomed for me in autumn 2008. Like *C. wattiorum*, *C. nerimaniae* is a plant of low elevations (700 m) and therefore requires similar growing conditions. I have only grown it in pots.

Crocus kotschyanus K. Koch

Crocus kotschyanus as well as the previously mentioned *C. speciosus* seem to be among the most widely grown autumn-flowering crocuses. Unfortunately, the reputation of *C. kotschyanus* is somewhat muddled, because some Dutch companies distribute an excellently growing and multiplying clone which either never blooms or forms very few malformed (four-petalled) flowers. For some time it was almost impossible to buy well-flowering plants. Now the situation has improved, but not that long ago, in the autumn of 2007, I again bought 100 corms of this clone under the name *C. kotschyanus* 'Albus', which means that it is still being grown by some nurseries. I don't know whether the non-blooming, malformed plants represent a genetic issue or a virus infection, but in any case I grow crocuses only for flowers and not for forage.

In my younger years *Crocus kotschyanus* was much better known as *C. zonatus*, a name which more precisely describes this species. Near the base of each tepal are two yellow spots that in some forms blend and merge, forming a bright yellow zone around the flower's throat. Unfortunately, in accordance with the rules of the *International Code of Botanical Nomenclature*, the priority belongs to the name *C. kotschyanus*, which is quite difficult to pronounce. The name honors botanist Theodor Kotschy who, in the nineteenth century, traveled a great deal through Turkey.

Corms of *Crocus kotschyanus* are somewhat flattened, never particularly symmetrical, and usually very large. Normally they increase by small rice-grain sized cormlets produced abundantly around the large corms; some of them are the size of an ant's egg. They detach from the mother corm so easily that it is very difficult to collect them all and so this species can become a quite unpleasant weed in crocus beds as almost all of the cormlets germinate. Recently populations were discovered (HKEP 9317) that increase by producing white stolons from the mother corms, forming small cormlets at their ends. This unusual trait for crocuses has till now been known in only a few species.

In the wild *Crocus kotschyanus* is very widespread. Brian Mathew divided it into four subspecies, which are quite easy to tell apart. One of them is now regarded as *C. suworowianus*.

Crocus kotschyanus subsp. *kotschyanus*

PLATES 31–33

Crocus kotschyanus subsp. *kotschyanus* ($2n = 8, 10$) is recorded in central and southern Turkey entering northern Syria and some localities in northeastern Lebanon. It is found in mountain meadows where it grows on open stony places and in sparse scrub from 550 to 2600 meters. The corms lie upright in the ground.

The flowers of subsp. *kotschyanus* are fragrant, generally pale to bluish lilac with some more or less pronounced darker lilac stripes that are more prominent at the base, and with a creamy white throat that is surrounded by very bright yellow basal blotches. The filaments are creamy and the anthers creamy white to white. The style usually overtops the anthers, rarely is equal to them, and is creamy yellow to orange. The style in Turkish plants is orange-yellow and divided into three branches, which are subdivided into several shorter ones.

Helmut Kerndorff (1994–1995) observed a fairly uniform population of subsp. *kotschyanus* in northwestern Syria in which all individuals had pure white styles rarely dissected into more than three branches. In 1992 H. Kerndorff and Erich Pasche in the province of Hatay in Turkey near the Syrian border found a population (HKEP-9205) developing long leaves in autumn. With me the leaf length of this sample reaches up to 8 cm before frost starts. Maybe it is caused by the fact that this form comes from a rather southern position, where it is warmer than in the rest of the country.

A form with a white throat is distributed under the name var. *leucopharynx* (Plate 33). The throat in this form is pure white without any yellow spots, but otherwise the shape of the flower's white throat is like that of the typical form of *C. kotschyanus*—with an incurved middle part. For many years this variety was grown and offered in catalogs under the wrong name of *C. karduchorum*. It was reported as being collected somewhere in Cilicia but was never rediscovered. As it only rarely sets seed, it is possible that in reality it is a clone. The corms are very misshapen and produce shoots somewhat asymmetrically; the shoots lay flat on the upper side of the corm. It is better to wait a little before replanting the corm, as the young shoots will show you the correct spacing for the corm in the soil. Var. *leucopharynx* blooms somewhat later than the type *C. kotschyanus*, and its flowers are a bit more bluish compared to the generally lilac-colored forms of typical plants. It is one of the fastest increasing crocuses—on average one large corm each year produces 3.5 new ones; the maximum has been 5 new flowering-sized corms.

David Stephens (1996) writes that from time to time similarly colored specimens appear among seedlings of the ordinary subsp. *kotschyanus* and in this way the variety originated—as a routine mutation. Now there are several forms to which the name var. *leucopharynx* can be applied. One of the best I saw in a picture from Thomas Huber (Germany); it had almost white flower segments with a round creamy white throat surrounded by lilac radial stripes, wider at the base and becoming very narrow near the tips. The anthers and style in this form are pure white.

An excellent purest milk-white form of *Crocus kotschyanus* has been selected and named **'Albus'**. The throat is cream-colored surrounded by an almost uninterrupted narrow bright deep yellow zone. I have some concerns about the stocks of 'Albus' which are offered now; very likely they are mostly virus-infected as the flowers do not open freely and the tepal surface seems a little crumbled.

The type subspecies of *Crocus kotschyanus* is very easy in the garden and can even naturalize. In my experience it doesn't increase by self-sowing (although it sets seed abundantly), but some unharvested bulbs plowed deep in the old crocus field have continued to bloom for more than 10 years. If you grow this species in pots, it is better to continue watering until the leaves start to yellow; during the summer rest period, the pots must be moved outside. Due to the high rate of increase (on average 3.1), an annual repotting is advisable.

In recent years in the trade a crocus has appeared that is offered as a hybrid between *Crocus kotschyanus* and *C. ochroleucus*. It is lighter in color than the mother species, but retains the very bright yellow basal blotches and flowers very abundantly in such a way supporting its hybrid origin. Its throat is deeper yellow than in the true *C. kotschyanus* but is similarly surrounded by a very prominent deep yellow zone. In *C. ochroleucus* it is uniformly deep yellow. Its leaves appear only in spring, so it looks more like *C. kotschyanus* than *C. ochroleucus*. The corm tunic is similar to that of *C. kotschyanus* but the corm itself is smaller.

In all the other subspecies of *Crocus kotschyanus* the corms in soil lie on their side. Both subsp. *cappadocicus* and subsp. *hakkariensis* are described as generally lilac colored though the flower segments of the samples that I have seen all had white background with more or less striking lilac veins. *Crocus suworowianus*, earlier regarded as a subspecies of *C. kotschyanus*, in general has white flowers.

Crocus kotschyanus* subsp. *cappadocicus B. Mathew

PLATES 16 & 34

Crocus kotschyanus subsp. *cappadocicus* ($2n = 10$) occurs in the central part of Anatolia (named after Cappadocia) and in some places is quite wide-spread on open rocky slopes and in short grassland from 1980 to 2700 meters.

In general the yellow spots at the tepal base are lighter yellow than in the type subspecies, but in one sample grown from seed received from Jim and Jenny Archibald they are of almost the same brightness. Similar quite dark spots at the tepal inner base are seen in samples collected by our team in short grassland near the Ziyaret pass close to the *locus classicus* (BATM-372, LST-364). The color of the base of the flower segments is whitish but densely veined lilac, and in some plants the veining at the top of the tepals merges and flushes, completely hiding the whitish background. This subspecies has very distinct leaves—they spread flat on the ground and are dark green. In other subspecies the leaves are brighter green and more erect. Growing in more open and rockier places are *Crocus biflorus* and *C. cancellatus*, while lower places with tall grass are occupied by *C. kotschyanus* subsp. *cappadocicus*. In spring the soil there is very wet, even overflooded, but in summer it dries up to brick hard.

Subspecies *cappadocicus* is a very beautiful plant that increases well and readily sets seed. Even when open-pollinated, the seed reproduces perfectly the type plants. I grew plants of this subspecies quite successfully for many years in the open garden but now have moved them to the greenhouse. The corms must be positioned on their side at planting.

Crocus kotschyanus subsp. *hakkariensis* B. Mathew

PLATE 35

I have only one sample of *Crocus kotschyanus* subsp. *hakkariensis* ($2n = 10$), which I obtained from a Czech collector as an unknown *Crocus* sp. from Turkey. In general appearance my plants are even more whitish than those of subsp. *cappadocicus*. The veining on the tepals is somewhat more purplish in shade but not so intense and does not merge at the tepal tips. The basal blotches are very light yellow, but because my knowledge of this subspecies comes from this single collection only, I admit that there could be variations in this aspect.

A very special feature of subsp. *hakkariensis* is the shape of the flower segments which are distinctly wedgelike and somewhat angular at the tips. In some aspects the flower looks slightly similar to some forms of the very variable *Crocus suworowianus*. I'm still not very certain about my identification as I don't know exactly where my plants were collected and according to Brian Mathew its flowers must be pale lilac and not very prominently veined. Just the shape of the tepals and the pubescent throat which is glabrous in *C. suworowianus* allows me to separate it from all the other subspecies of *C. kotschyanus*.

In the wild, subspecies *hakkariensis* is known only from the province of Hakkâri in southeastern Turkey where it grows on rocky or gravelly places in mountain steppe vegetation from 1100 to 3250 meters, far from the main distribution area of *Crocus kotschyanus*. It needs hot and dry conditions in the summer and therefore is suitable only for pot growing. My plants set seed well, and the seed germinates readily. Corms must be planted on their side.

Crocus suworowianus K. Koch

PLATES 10, 17 & 36

Crocus suworowianus ($2n = 20$) is described from South Ossetia and occurs in northeastern Turkey where it grows from 2000 to 3100 meters. In the World Checklist of crocus names the name is spelled as *suw**a**rowianus*, but as the species is named after Prince Konstantin Suvorov, the spelling has to be corrected to *suw**o**rowianus*. Likewise, *C. tommasinianus* originally was written with one *m*, but botanist Muzio de Tommasini, after whom it was named, always spelled his name with two *m*'s, so the name of the species must be spelled with two *m*'s, as it usually is now written.

In typical forms *Crocus suworowianus* is not very difficult to identify. The flowers are white or rarely pale lilac with a glabrous throat which allows separating this species from *C. kotschyanus* subsp. *cappadocicus* in which the flower segments are intensively striped. Another lookalike is *C. vallicola*, but it has a very distinct tepal shape and its corms are oriented upright in the soil, not on the side as is typical for *C. kotschyanus* subsp. *cappadocicus* and subsp. *hakkariensis* (the latter has a pubescent throat) and for *C. suworowianus*. Nevertheless, it is not always easy to identify this species. I obtained several samples collected by Czech travelers under the name of *C. suworowianus* var. *lilacinus*; these had a glabrous throat but the tepal color was almost identical or even more lilac than that of *C. kotschyanus* subsp. *cappadocicus*. Only the very pale basal blotches allow me to grow them still under the name of *C. suworowianus*.

I collected one plant of *Crocus suworowianus* on Yaylasuyun mountain pass at an elevation of 2330 to 2400 meters (BATM-200). It has pure white flowers, but the tepals are distinctly wedge-shaped as in *C. kotschyanus* subsp. *hakkariensis*. Another plant from the same locality has acuminate tepal tips, somewhat resembling those of *C. vallicola*. Plants from a sample from the Kisidagi pass collected at an elevation of 2120 meters (BATM-321) have lilac flowers and the basal blotches are light to deep yellow. All of these forms have corms that lie sideways in the soil, otherwise I would prefer the name *C. kotschyanus* subsp. *kotschyanus*. In some forms the base of the flower segments is quite distinctly striped lilac. Our team collected such a sample in the Kackar Mountains above the village of Barhal (LST-304).

Some Russian botanists (Takhtajan 2006) regard *Crocus suworowianus* as a hybrid between *C. vallicola* and *C. scharojanii*. Such hybrids exist but they are pale yellow and their accepted name is *C. scharojanii* var. *flavus*. More likely some hybridization could occur between *C. suworowianus* and *C. kotschyanus*, but they don't meet in the wild.

Crocus suworowianus comes from high mountain meadows where it grows in dryish grassland or in rocky places in upland steppe. It is not very difficult in cultivation, but because I have lost it several times when trying to grow it outside, I recommend pot growing only. It needs somewhat drier growing conditions and the flowering starts a little later than with *C. vallicola*. During the summer I keep the pots outside where it is cooler and they get occasional rain. The corms must be planted on their sides.

Crocus vallicola Herbert

PLATE 37

In general appearance *Crocus vallicola* ($2n = 8$) is closest to *C. suworowianus*. The most distinct feature which easily allows separating it from its cousins is the very long wispy tips of the perianth segments. It is a plant from high valleys of northeastern Turkey and adjoining parts of Russia up to the west of the Caucasus ridge, where it grows on granite formations in open subalpine pastures and alpine grassland from 1000 to 3000 meters. Its name means "valley dweller," but that doesn't signify that it grows in valleys. Usually it is found at higher elevations in places somewhat depressed where more moisture is preserved during the entire summer, but at the same time the soil isn't damp. This crocus prefers slightly drier grassland than its close relative *C. scharojanii*. On the Zigana pass it grows side by side with *Colchicum speciosum*. It is one of the earliest autumn-blooming crocuses.

My first samples of *Crocus vallicola* were originally collected in the Russian Caucasus at the famous ski resort of Dombai, Stavropol district, by Aino Paivel. I got them from the Tallinn Botanic Garden, where plants grew perfectly in an outside garden. I grew the species for many years in a somewhat shaded spot in my garden, and it multiplied vegetatively and set many seeds until a very harsh winter when I lost all my Russian vallicolas. All that remains from the Dombai plants is a few slides and detailed notes made in my garden where the plants regularly produced flower buds on the 15th of August and bloomed from the 20th of August onwards. There are forms described from Dombai (Kos 1948) with very densely veined tepal insides and, as a rule, more intense veining is joined with darker spots (up to orange) at the tepal base.

Usually at the inside base of each *Crocus vallicola* flower segment are a couple of pale yellow spots, and they are even paler than in *C. suworowianus* though I have several samples of *C. vallicola* in which these spots are completely absent. Sometimes they can be so large that they make an almost complete ring. The purple veins which surround the throat can be completely or nearly absent, clearly developed or set off by extra feathering on the outside usually reaching half the tepals length, but in some samples from the Zigana pass they almost reach the tepal tips. The veining varies much among plants in the same population. The most interesting one comes from Artvin where it was collected by Milan Prasil (MP 8812); the upper rim of its perianth segments looks somewhat hairy or slightly fringed, but the wispy tips of tepals are absent. The color of the style branches varies from cream to deep yellow.

The Turkish forms of *Crocus vallicola* are not of such refined beauty (with a few exceptions) as those from Dombai, and I grow them here without any problems. The largest part of my stock is now pot-grown, and I regularly check the soil to be sure it does not dry out completely. I keep the lifted bulbs of this species in open boxes covered with silvery sand, but it is very important to start replanting early for they commence forming flowering shoots in the beginning of August here. This means they can start flowering in boxes if replanting is delayed. For *C. vallicola* I use a soil mix that is slightly richer in humus and add a small amount of peat moss. It increases moderately by vegetative means and sets abundant seed as its blooming happens when the weather is still warm and pollinators are active.

Crocus scharojanii Ruprecht

PLATES 38–40

A close relative of *Crocus vallicola* but much more difficult in the garden is *Crocus scharojanii* ($2n = 8$). I grew *C. scharojanii* for many years in the open garden without many problems until one winter when water rats cleaned the entire stock from my collection. I had planted a 10-meter-long bed with two rows of *C. scharojanii* and not a single corm was left! It was one of the occasions when I had forgotten the rule: never put all your eggs in the same basket, that is, never plant all your corms of a particular crocus in one spot.

Crocus scharojanii is the only autumn-flowering crocus with yellow blooms, although it should be better called late summer-flowering crocus. Here its blooming usually starts in the last two weeks of July and ends in the second half of August. The plants almost always flower before the old leaves die; at least the new roots start to grow long before the old ones have withered away so there is no real optimal time for replanting.

In the wild this species usually replaces *Crocus vallicola* at higher elevations and in much wetter conditions, being a plant from wet peaty grassland near rivulets and depressions where the soil never gets dry. In some places the two species grow side by side and can hybridize. This crocus was named after Mr. Scharojan, who collected it in the Caucasus in 1865.

On the basis of the stoloniferous and non-stoloniferous habit, the shape of the tepals, and the size of corms, Brian Mathew (2002) recently divided *Crocus scharojanii* into two subspecies—the type ***C. scharojanii* subsp. *scharojanii*** from the Caucasus and ***C. scharojanii* subsp. *lazicus*** (former *C. lazicus* but merged with the type subspecies in his monograph) from northeastern Turkey.

I grow several stocks originally collected in Turkey (subsp. *lazicus*) which I obtained from various sources. I have never gotten seeds from these plants, even after hand pollinating them. The Turkish plants now have to be named subsp. *lazicus*. Following Brian Mathew the Turkish plants have smaller corms and sometimes produce stolons—a feature not observed in the Caucasian plants (subsp. *scharojanii*), but I have never observed stolons in my Turkish plants either. Studying the pictures I have, I can see that the tepals of the Caucasian forms (type subspecies) are more pointed and more yellow than the Turkish forms, which are more rounded and a brighter orange-yellow color. The corms of the Caucasian forms are larger than those of the Turkish forms and are thinly membranous with the fibers parallel at the base and slightly reticulated at the top.

I again got some corms of Caucasian subspecies from different locations only very recently and haven't had an opportunity to compare both subspecies for myself. My Ukrainian friend Dmitriy Zubov has photographed a beautiful creamy white individual of the type subspecies between millions of bright yellow flowers near Great Dukka lakes in the northern Caucasus at an altitude of 1900 m.

Brian Mathew (2002) writes that the Caucasian forms are easier in cultivation than the Turkish ones, and that in the wild they grow in somewhat drier conditions. All my samples of *Crocus scharojanii* are grown in pots and I replant them once every two or three years. The soil mix has an additional portion of peat moss. During the summer I put the pots outside, where the soil is really wet, until quite late in autumn when I stop watering and keep the pots slightly drier during the frost season.

Crocus scharojanii* var. *flavus Lipsky

PLATE 41

A somewhat easier grower is *Crocus scharojanii* var. *flavus*, also known as *C. vallicola* var. *intermedia* Yabrova. Its flowers are a very beautiful light lemon yellow color. This crocus is a hybrid between *C. scharojanii* and *C. vallicola*, although some gardeners tend to think that it is a different species. The hybrid origin was confirmed by cytological studies. From *C. vallicola* it has inherited the long acuminate tips of the flower segments. It is more floriferous and somewhat easier to grow, increases better, and sometimes sets seeds. In my experience, var. *flavus* blooms later than *C. scharojanii*. I keep it in the same conditions as *C. scharojanii*.

Crocus autranii Albov

PLATES 42 & 43

The last Caucasian autumn-flowering crocus is the very endemic growing *Crocus autranii* ($2n = 32$). It was named by Swiss botanist Eugene Autran, Keeper of the Boissier Herbarium in Geneva. It is known only from Abkhazia where it grows in alpine meadows at 2300 meters. According to *Flora of Abchasia* (Kolakovski 1986), it is quite common on calkstone-based hills. Later in *Caucasian Flora Conspectus* (Takhtajan 2006) it is characterized as extremely rare and almost unknown, and the idea is expressed that it is possibly only an abnormal form of *C. speciosus*. That certainly isn't true. On the phylogenetic tree it is placed very far from *C. speciosus* and side by side with *C. gilanicus* and *C. scharojanii*.

The flower shape is somewhat intermediate between that of *Crocus vallicola* and *C. suworowianus*. The perianth segments are wider and less sharply pointed than in *C. vallicola*, but the flower color is rich midviolet with dark parallel veins along the length of the segment. The flower throat is large pure white without any traces of yellow spots at the base, rarely with very pale spots. The filaments and anthers are white; style branches are orange and overtop the anthers.

My only attempt to find this species in the mountains in the early 1980s was interrupted by a sudden snowfall that covered all the meadows with a thick layer of snow and forced me to return to Sochi and back home before I managed to see at least one single flower. Subsequently it was collected by a Czech plant hunter and from there the species found its way into gardens, but it is still extremely rare. My first corm came from the Gothenburg Botanical Garden. After it produced a flower, which I hand-pollinated, it set one seed and the seedling bloomed in its fourth autumn, allowing me to cross-pollinate it with the mother plant and obtain a far better seed crop.

Crocus autranii is comparatively easy to grow here in the same way as *C. vallicola* only it seems to be more susceptible to winter frosts. I grow it in large pots which overwinter in my cold greenhouse. As soon as the weather permits, I move the pots into the garden and constantly check that the soil never dries out. It goes without saying that excellent drainage is a must. When pollinated artificially, this species sets seeds readily. It can be crossed with *C. gilanicus*. First-generation seedlings are very uniform and look very similar to *C. autranii* except they are slightly lighter in shade. Hybrids are fertile and set seeds but second-generation seedlings have not yet flowered for me.

Crocus karduchorum Kotschy ex Maw

PLATE 44

Another leafless autumn-bloomer with a pure white throat is *Crocus karduchorum* ($2n = 10$) from southeastern Turkey where it grows south of Lake Van. The area was earlier known as Carduchi and is the source of the species name. In the wild this species grows from 1850 to 2000 meters, on steep banks between oak trees and scrub on gritty non-calcareous soils which dry out in summer.

Crocus karduchorum is quite new in gardens although it was discovered in 1859. The name was well-known to gardeners, but for many decades the plant offered by nurseries under this name was *C. kotschyanus* var. *leucopharynx*, which has a similar pure white throat but a very different style. The true *C. karduchorum* was rediscovered in the wild only in 1974 by Turhan Baytop.

This species is very easily distinguished from other crocuses. It has multiple strikingly white style branches that overtop the anthers and almost reach the tips of the flower segments. The white base is diffused and gradually blends into the lilac tepals, and the corms lie in the ground on their side. In contrast, the flowers of *Crocus kotschyanus* var. *leucopharynx* are a darker lilac color, the white base is sharply defined from the lilac tepals, the style is much less branched and does not overtop the anthers.

In my collection is a plant named *Crocus karduchorum* and, although it looks like the true *C. karduchorum*, at the tepal base are pale yellow spots, the style is less branched, creamy yellow, and reaches only half the tepal length although it well overtops the anthers of the same color. In typical *C. karduchorum* the anthers and filaments are pure white. Probably my plant is a hybrid with *C. kotschyanus* as some of its forms have an identical chromosome number ($2n = 10$) with *C. karduchorum*.

My stock is still too small for experiments in the open garden although, before I put up my greenhouses, I for several years grew it in open beds. It seems to be somewhat less hardy than *Crocus kotschyanus*. Pots must be kept dry in summer.

Crocus gilanicus B. Mathew

PLATE 45

One more species with thinly membranous tunics and corms lying on their side or at least oblique in the ground grows quite far to the east—in Iranian Azerbaijan, in the provinces of Azerbaijan and Gīlān after which it is named. *Crocus gilanicus* ($2n = 24$) occurs in beech forests and grazed meadows from 1500 to 2400 meters. It was found in the wild for the first time only in 1973 by A. Shirdelpur from the Ariamehr Botanic Garden in Tehran. Shirdelpur found it in pastures above the treeline in the southern Talish Mountains at 2400 meters. Phylogenetic studies place it side by side with *C. vallicola*, *C. scharojanii*, and *C. autranii*.

From a gardener's viewpoint, this crocus is a "poor relative" of the family as its flowers are comparatively pale and small. For me it mostly resembles *Crocus suworowianus* with its white flowers slightly veined lilac but without the yellow spots in the white throat which is pubescent. The throat of *C. suworowianus* is nude or glabrous. The anthers are of equal length to the filaments but in *C. suworowianus* they are two to three times longer. Both species differ genetically, too. Few plants of *C. gilanicus* are grown in cultivation and possibly more spectacular variants can be found in the wild. A "dark form," grown by Tony Goode, is more colored than the usual form; it is also smaller and quite different in shape. *Crocus gilanicus* can be crossed with *C. autranii* and such a hybrid was raised in the Gothenburg Botanical Garden.

Crocus gilanicus is not very hardy and once I lost it even in the greenhouse. I have never tried *C. gilanicus* outside, but in pots it is quite easy if you don't forget that it doesn't like complete drying out in summer. For this species I use my standard soil mix. Annual repotting and placing of corms on their side are essential. The dormant corms should not be allowed to dry out. Foliage appears late in spring, usually just when I start to worry about how well the corms have overwintered. The seed crop is irregular but the corm produces a lot of cormlets and vegetative increase is quick.

Crocus nudiflorus J. E. Smith

PLATE 46

All the other leafless autumn-flowering species you can easily separate by their much coarser fibrous tunics. Among them is another species with a stoloniferous habit, namely, *Crocus nudiflorus* ($2n = 48$), which comes from the western corner of the genus's distribution—southwestern France and northern and eastern Spain. It occurs from sea level up to 2000 meters, in open moist meadows where a substantial amount of rain falls in summer. Its name clearly indicates that the plant is leafless at the flowering time. This species has naturalized in England and the type sample used for the description came from such a naturalized population.

The flowers of *Crocus nudiflorus* in most cases are deep purple though sometimes albinos can be found with or without the purple staining in the throat. The throat in all forms is invariably white and hairless, whereas in other Spanish crocuses it is more or less pubescent and they develop leaves at least during the flowering if not earlier. In very rare cases the leaves appear just at the end of blooming, but then the hairless and colorless throat of *C. nudiflorus* helps in identification. Its corm tunic is membranous interspersed with strong parallel fibers; the older tunics become more or less wholly fibrous. The stoloniferous habit, which is rare in *Crocus*, can also be seen in some Spanish populations of *C. serotinus* subsp. *salzmannii*.

My first corms of *Crocus nudiflorus* came from Michael Hoog. I was surprised by the unusual shape of stoloniferous bulblets included in the parcel. Sometimes they are very slender, other times they resemble the roots of the couch-grass only shorter in length, and sometimes they look like an enormously extended corm (Fig. 4). E. A. Bowles wrote that "anyone seeing them for the first time might be pardoned for mistaking them for some evil form of stoloniferous grass." When planted in the garden, the stolonlike corms take two or three years to form flowers. This species has turned out to be a good grower here, and I have never lost it completely even after the harshest winters. When well-grown it is quite attractive and when planted in a rock garden it slowly spreads out by stolons. Although George Maw (1886) characterizes it as "one of the largest and most ornamental species of the genus, and one which no garden should be without," this species is not one of my favorites. Perhaps if I had obtained a pure white form or even the more "white variety with a purple throat and stripe" mentioned by Maw, I would think otherwise.

In regions with a milder climate than that of Latvia *Crocus nudiflorus* can grow in grass as well. It sets seed readily, and the seedlings bloom in the third or fourth season after sowing. Its main requirement is not to be disturbed in order to be able to spread and multiply by stolons. If it is planted in pots they must be kept outside during the summer and some care must be taken to avoid overdrying.

Crocus robertianus C. D. Brickell

PLATES 47 & 48

But now let's return to the eastern direction of the crocus range. In northern Greece, scattered here and there throughout the large Pindus range, in deciduous oak and juniper scrub on sandstone shale formations from 450 to 900 meters, you can find one of the largest autumn-blooming crocuses, namely, *Crocus robertianus* ($2n = 20$). In flower size, it and *C. niveus* are the largest crocuses from the Greek mainland.

Crocus robertianus was collected for the first time only in 1967 by J. R. Marr and named after his son Robert, who died in early childhood, by Chris Brickell in 1973. Later it was found in beech woods as well, but it seems to favor dappled woodland edges.

Crocus robertianus has a strongly reticulated corm tunic. So does *C. cancellatus*, but the style of the latter is many-branched, so there is no possibility of misidentification. *Crocus robertianus* is one of the best autumn-flowering crocuses, with large light lilac-blue flowers and a very bright orange-yellow throat edged with a diffused yellowish zone. A large bright orange to red frilly stigma well overtops the creamy to bright yellow anthers giving additional beauty to the wide-tepalled flower of excellent form. The flower color varies slightly in the depth of the shade and sometimes pure white or only slightly lilac-shaded plants can be found. In their beauty and color they can compete with the spring-blooming *C. atticus*. Brian Mathew thinks that *C. robertianus* has evolved by mutation from *C. sieberi* (here regarded as *C. atticus*). In the phylogenetic tree the two species are placed close to each other.

Generally *Crocus robertianus* grows in clearings and at the edge of scrub and forests where some shade is provided and the soil is moist from autumn till late spring but kept dry by the tree roots in summer. In my experience, this species is less hardy than its close relative *C. atticus*. While cultivating *C. robertianus* in the open garden I lost it several times. It bloomed only in years with long, warm autumns, usually in the second half of November, but here permanent frost can set in even in the end of October. It grows well in pots but has to be kept in a slightly shaded place during the summer so the pots do not get too much heat and do not dry out excessively. My favorite form is the white-flowered variant which in beauty is second only to its spring-blooming relative *C. atticus* 'Bowles' White'.

Crocus cancellatus Herbert

Crocus cancellatus is another well-known autumn-blooming species with a corm tunic similar to that of *C. robertianus*. The name means "latticed," and the corm tunics truly are the most coarsely netted in the whole genus. Among the autumn-blooming crocuses the only competitors in tunic characters are *C. robertianus* which has a trifid stigma and *C. ligusticus* (more widely known as *C. medius*) which is easily distinguished by its green bract. *Crocus cancellatus* is widely distributed from southern Yugoslavia and Greece through Turkey as far as Iran in the east and Jordan in the south. Given this large area, it is no wonder that five subspecies have been recognized by Brian Mathew.

Crocus cancellatus sensu lato has no close affinities with other autumnal species. The flower color varies from white to rich lilac and the outer tepals are diversely feathered or striped with purple. My favorite plants in terms of color are bluish lilac with a white throat extending like white stripes up half the length of the tepals. Sometimes it isn't easy to decide—are these white stripes on a purplish ground or are they lilac stripes on a white ground? The tunic is coarsely reticulated, the stigma many-branched. Leaves appear after the flowers, but in some forms the green leaf tips can appear with the wilting of the last bloom.

Of the five subspecies, one is easy to recognize because it has white anthers. The remaining subspecies are a bit more difficult to separate. They can be put into two pairs by the position of the stigmatic branches relative to the anthers and by whether the bracteole is visible or hidden within the bract.

Crocus cancellatus* subsp. *pamphylicus B. Mathew

PLATE 49

The easiest subspecies to recognize is *Crocus cancellatus* subsp. *pamphylicus* ($2n = 12$). It differs from the other subspecies primarily in the white color of the anthers. It grows in a restricted area in southern Turkey—in the provinces of Antalya and West Içel, the classical region of Pamphylia in the western Taurus after which it is named. It occurs from 250 to 1500 meters in rocky places in open conifer woods and scrub.

The flower color in most of my samples is almost white with a few short narrow purple veins at the tepal outer base extending down to the flower tube, although it can be pale lilac as well. The throat and filaments in most cases are deep yellow, the throat sometimes being orange. In my samples the style has fewer divisions than in other subspecies, but this feature is very variable. Since its natural habitat is an area with high rainfall, it must be more tolerant to wetter growing conditions than the subspecies growing further east. I grow it only in pots, which I keep dry and warm during the summer rest.

Crocus cancellatus* subsp. *mazziaricus (Herbert) B. Mathew

PLATE 50

Crocus cancellatus subsp. *mazziaricus* ($2n = 16, 18$) is the most westerly growing crocus in this complex. It is very widespread in Greece and enters southern Yugoslavia as well as western and southwestern Turkey, growing on rocky hillsides in open woods or in scrub on limestone formations, often in terra rosa from sea level to 1500 meters. It was named after A. D. Mazziari, who originally collected it, and was regarded as a species by William Herbert, who described it.

The flower color can vary widely from white to pale lilac to rich bluish lilac. The plant is very common in many Greek mountains. All the plants of subsp. *mazziaricus* seen on Mount Chelmos had white flowers. Very nice blue forms were seen by Kees Jan van Zwienen in the foothills of Mount Pelion, eastern Greece. The darker-colored forms are more common in the northern part of its area while the lighter ones come from Peloponnesus and southwestern Turkey. From the Honaz-Dag (Turkey) are known specimens with a purple throat. The many-branched style greatly exceeds the anthers. Sometimes the style can be almost undivided and if this feature is combined with a purple throat, then only strongly reticulated corm tunics allow separating it from *Crocus mathewii*. Although the bract and bracteole are unequal in size, the tip of the bracteole is visible, exserted from the tubular bract.

This subspecies is the most ornamental one in the complex. I would have preferred to see it keep its species status as it differs from the other subspecies in chromosome number. It partly merges with subsp. *lycius* in which are recognized forms with $2n = 16$.

I grow my samples only in pots, ensuring them a dry and warm summer rest.

Crocus cancellatus* subsp. *lycius B. Mathew

PLATE 51

Another subspecies in which both the bract and bracteole are visible is *Crocus cancellatus* subsp. *lycius* ($2n = 14, 16$). Brian Mathew separated it for its widely spread and much-divided style branches which are equal or shorter than the anthers. It grows only in southwestern Turkey (in Lycian Taurus) where its area partly merges with that of subsp. *mazziaricus*. It occurs on rocky slopes of limestone formations often in maquis (a type of shrubland) and sometimes in grazed grassland from 300 to 1400 meters.

Flowers in all my samples are pure white with a light yellow to straw-colored—sometimes even deep yellow—throat, and they are smaller than flowers of subsp. *mazziaricus*. The greatest "bonus" is the bright red or deep orange stigmatic branches which are so divided that they curve around the anthers, filling the central part of the flower with a beautiful dance of colors. Occasionally the white flowers are faintly flushed lilac. It is so special and distinct that I would like to regard it at species level.

Like all the other subspecies of *Crocus cancellatus*, this one can be grown only in a greenhouse where a dry and warm summer rest can be provided.

Crocus cancellatus* subsp. *damascenus (Herbert) B. Mathew

PLATE 52

At the easternmost end of the complex's range is *Crocus cancellatus* subsp. *damascenus* ($2n = 8, 10, 12$), which William Herbert regarded as a separate species *C. damascenus*. In this subspecies the bracteole is present but, as in the type subsp. *cancellatus*, it is hidden by the bract and not usually visible without dissecting the flower. The subspecies is widely distributed from Jordan in the south up to northern Iraq, western Iran, and central, eastern and southeastern Turkey. It grows in open stony places, dryish fields, and clearings in oak scrub on limestone or basaltic formations from 300 to 3500 meters.

In the Negev Desert in Israel, subsp. *damascenus* in some years flowers two to three months before the autumn rains start, while daytime temperatures are still above 30°C. The peak flowering is in mid-November, as it is for most flowering desert species, and this crocus doesn't depend on the first rains in order to flower (Oron Peri).

The style branches of subsp. *damascenus* are often overtopped by the anthers. The degree of stigmatic branching can vary from few (but always more than three) to many. The flowers are pale to midlilac-blue, often with darker stripes on the sepal outside base. The throat color varies from grayish white to very pale yellow. Subsp. *damascenus* can be easily identified by its very coarse tunic netting—the coarsest in the whole complex—forming a bristly fibrous neck up to 6 cm long and with interspaces of 6 mm or more. It requires the hottest and driest summer rest.

Crocus cancellatus subsp. *cancellatus*

PLATES 53 & 54

The type *Crocus cancellatus* subsp. *cancellatus* (2*n* = 10, 12) is the only subspecies that can be readily bought from bulb companies as it is fairly widely grown by the large nurseries in Holland. In the wild, subsp. *cancellatus* grows at elevations from 50 to 2400 meters on open stony slopes and in sparse pine woodlands on limestone formations in southern Turkey, western Syria, Lebanon, and northern Israel. According to the phylogenetic tree of the genus, the closest ally of this subspecies is subsp. *pamphylicus*, while the other three subspecies are placed side by side but at some distance from the type subspecies.

Compared with the corm tunics of subsp. *damascenus*, those of subsp. *cancellatus* are more finely netted with interspaces not exceeding 5 mm, and the bristly fibrous neck is very short. The many-branched orange style usually exceeds the length of the anthers, although sometimes it can be the same length. The flowers generally are pale to midlilac blue, usually with violet veins, on the outside merging at the tepal base. The throat is whitish to light yellow.

The veining is especially prominent in the commercially distributed *Crocus cancellatus* var. *cilicicus* (Plate 54), in which the deep purple stripes on the exterior of the outer segments reach the tips of the sepals. The throat of var. *cilicicus* has purple stripes crossing the greenish yellow zone on the inside of the tepal and moving on up to the tepal tips. This variety is the easiest form of the complex in cultivation. I have grown it with varying success in the outside garden and cannot recall that I ever lost it completely, but it grows far better in a pot under cover. I have never tried to grow other subspecies of *C. cancellatus* in the open.

Crocus ligusticus Mariotti

PLATE 55

Crocus ligusticus (2*n* = 24), which in gardens is better known under the name of *C. medius*, is easily distinguished by its green bract which is uncolored in other autumn-blooming leafless crocuses. In the wild this species grows in meadows on mountain pastures and in woodland in southeastern France and northern Italy from 200 to 1400 meters. According to the phylogenetic tree, its closest ally is *C. goulimyi* from Greece. The name was changed after a careful check of the herbarium sheet used as the type specimen for *C. medius* turned out to show *C. nudiflorus*. The next earliest epithet of the true plant, in this case *C. ligusticus*, became the new, correct name.

Crocus ligusticus is a very handsome and comparatively hardy species usually offered in every bulb catalog (although still under the name *C. medius*). The greatest problem is that most commercial stocks are almost 100 percent virus-infected. Recently I received a marvelous plant from the German crocus enthusiast Thomas Huber, who presented me with a small corm collected on Mount Biguone in northern Italy. Despite its small size, this corm bloom marvelously for me in its first autumn. Tony Goode found that this crocus is very virus prone or perhaps it is just very quick to show the symptoms.

The flower in the best forms is a rich, deep lilac with nice dark stripes in the throat and a bright red many-branched style reaching almost the tips of tepals above the much shorter bright yellow anthers. There are records of pale-colored forms and even rare albinos have been seen in the wild. The flowers can vary greatly in size. The tepals of my plants usually are up to 4.0 to 4.5 cm long, but Alan Edwards showed me herbarium blooms of plants collected in the Alpes Maritime of southern France; these exceeded the size of the giant Dutch *C.* ×*cultorum* cultivars.

Crocus ligusticus can be grown in open garden here but for now I don't want to risk losing such a long-searched-for treasure. Therefore my few corms are in a pot which I take outside the greenhouse in summer.

Crocus hermoneus Kotschy ex Maw

PLATES 56 & 57

I have no experience with this leafless autumn-blooming species. *Crocus hermoneus* is distributed in central Israel and western Jordan. Although it was first collected near the summit of Mount Hermon in 1855, it is almost unknown in gardens. The plant needs a dry and hot summer rest and protection for the late-emerging flowers. I assume that the conditions in my nursery won't be acceptable for this species, therefore I have never searched for it. According to Brian Mathew *C. hermoneus* is "a poor plant compared to some forms of *C. cancellatus*." Its flowers are reported to be pale lilac-blue or sometimes white, usually with darker violet veins on the outer base and a white or pale lilac throat. The filaments are white, the anthers yellow and overtopping the six- (or more) branched yellow to orange style.

Crocus hermoneus can be separated from other species in this group by its fibrous tunics that split near the base into parallel fibers and at the top into weakly reticulated fine fibers which form a neck at the apex of the corm. Two subspecies are recognized. ***Crocus hermoneus* subsp. *hermoneus*** ($2n = 8$) grows at higher elevations and blooms earlier—in early and midautumn (September–October). Its corm tunic forms a neck up to 1.5 cm long, and it produces six or seven leaves 2.5 to 3 mm wide. ***Crocus hermoneus* subsp. *palaestinus*** Feinbrun ($2n = 12$) comes from lower elevations and blooms in late autumn and early winter (November–December). Its corm tunic forms a neck 2 to 3 cm long, and it produces three or four leaves up to 2 mm wide.

Crocus hermoneus was explored *en situ* by H. Kerndorff, who found that the differences between plants are not too significant and that they actually represent one somewhat variable species. It is normal in autumn-flowering species for plants occurring on the summits at higher elevations to start blooming while those growing at the lowest elevations bloom last. With spring-blooming species, the sequence is reversed. That leaves the question about the chromosome number. Still, there are several species in which samples with different chromosome numbers coexist. (Anyway, I included both subspecies in the identification key.) In some samples the corm tunics turned out to be very similar to those of *C. cancellatus*. As far as I can tell from the published pictures I've seen, the flowers of *C. hermoneus* very much resemble some of the *C. cancellatus* forms and it is almost impossible to distinguish them both without seeing the corms.

To my knowledge, *Crocus hermoneus* is only grown in some collections in the United Kingdom and Germany, in conditions somewhat imitating those in the wild. Helmut Kerndorff and Erich Pasche recommend giving the plants a dry and warm summer rest between May and October followed by plenty of water in autumn and in milder periods during the winter. The greatest danger to cultivating this species is *Botrytis*, which can be prevented by sufficient, continual air movement in the greenhouse. If good fertilization is provided with additional lighting, the resulting corms will be large.

Crocus moabiticus Bornmüller

PLATES 58–60

Another species which I have never grown is *Crocus moabiticus* ($2n = 14$), the only autumn-blooming member of series *Crocus* in which the leaves appear after blooming has ended. This species also differs from others in the series by its corm tunic, which in *C. moabiticus* is finely fibrous and only at the apex weakly reticulated, but in all the other species of the series is finely reticulated throughout.

Crocus moabiticus was discovered in the region of Moab in northern Jordan and described in 1912. Its flowers are starry, very pallid, being white with a few violet stripes; the throat is white, veined or suffused purple. The filaments are whitish, the anthers yellow, and the style trifid, which sometimes has two or three distinct lobes at the apex, reddish orange, equal to or slightly exceeding the anthers. According to Brian Mathew its flowers are "rather small, pale and starry so the species cannot be classed as a particularly garden-worthy subject." Mathew thinks that the species should be frost hardy even though it could suffer from long periods of dull, wet, cold weather in winter. *Crocus moabiticus* is one of three species in which once the flowers open, they never close until they fade.

In fact this species is very variable in the wild. The base color of the flower segments is white but mostly veined or more or less intensely feathered purple so that from a distance some flowers look uniformly violet-purple. Helmut Kerndorff has observed that the style arms are bright red and in some cases hang out of flowers and that the corm has a long neck. Flowering starts in autumn and continues until early spring. The number of leaves normally is 14 to 17 but can be as high as 24, making this species unique among crocuses. In cultivation the leaves emerge during flowering in November and December.

Oron Peri from Israel spotted *Crocus moabiticus* east of Madaba in Jordan and noted that "although *C. moabiticus* is very colorful, it is hard to notice it in its natural habitat, it somehow manages to blend very well with the color and structure of the soil." In this population corms usually produced one to three, rarely four flowers, mostly two. Like Kerndorff, Peri observed that the red style branches are often exserted from the flower in a circle.

Helmut Kerndorff and Erich Pasche report that plants suffer from the lack of light during winter months and so they provide additional lighting, giving approximately 2000 lux for 10 hours every day. Otherwise, *Crocus moabiticus* needs the same cultural treatment as that given *C. hermoneus*.

There are two other closely related species in series *Crocus* (*C. asumaniae* and *C. mathewii*) whose leaves usually are invisible at the beginning of flowering but they soon show their tips and are quite advanced by the end of flowering; therefore these crocuses are included with the autumnal crocuses that flower with leaves.

Autumnal Crocuses Flowering with Leaves

In this chapter we look at all the autumn-blooming crocuses that either form leaves before the flower buds appear or at least reveal leaf tips during the first half of blooming. These plants overwinter with at least partly developed leaves which continue growing in spring or, in milder climate, even during winter. Many of them can be grown outside in milder climates and some are grown in large numbers by Dutch nurseries as well as in many gardens in the United Kingdom. All my attempts to coax even the hardiest species in open beds failed in the first winter.

E. A Bowles divided all autumn-blooming species that produce flowers and leaves simultaneously into two groups. Those with ciliated margins on the blades and keel of the leaf he separated into "The Saffron Crocus" group. Here were included *Crocus sativus* and allied species, all of which belong to series *Crocus*. In the other group he left those species with smooth (glabrous) leaves. It would be easy to follow this classification, but unfortunately the leaf margin characters can vary greatly within the same species. For this reason, I don't think that a good identification key can be based on this feature.

I realize that grouping crocuses by the presence or absence of leaves at the time of flowering is a very artificial division. The key that follows is very long, and it covers species from seven different series. Several species are included more than once in the key. This is the case when the species is very variable. I have tried to cover some variability with the key, but, as I have emphasized previously, crocuses are so diverse that no one key can cover all the possibilities. A few species included in this key were discussed in the previous chapter on leafless autumn-blooming crocuses to avoid any misidentification in case of wider variation.

1. Corm tunics non-fibrous, smooth, splitting or with rings at base
  2. Anthers black or with black connective
    3. Flowers with white background, striped or speckled purple or gray on the exterior.. .... *C. melantherus*
    3. Flowers blue .... *C. wattiorum*
  2. Anthers yellow
    4. Tunics membranous, style well exceeding anthers .... *C. caspius*
    4. Tunics coriaceous
      5. Style trifid, shorter or slightly exceeding anthers .... *C. goulimyi*
      5. Style multifid, equaling or exceeding anthers .... *C. laevigatus*
1. Corm tunics membranous, splitting into many parallel fibers, not or only weakly reticulated at apex
  6. Anthers wholly or partly black (with black connective or margins)
    7. Leaves 2–2.5 mm wide with conspicuous wide white median stripe ...... *C. hyemalis*
    7. Leaves 0.5–1.5 mm wide with rather indistinct, narrow pale median stripe .... *C. aleppicus*
  6. Anthers white or yellow
    8. Style trifid
      9. Anthers white .... *C. ochroleucus*
      9. Anthers yellow
        10. Flowers white to lilac, with conspicuous purple stripes and veins on the outer segments, throat white, leaves 3–5.......... *C. cambessedesii*
        10. Flowers white with purplish veins, throat white, leaves 11–13
          11. Corm tunics with 5–6 cm long neck.......... *C. moabiticus*
          11. Corm tunics reduced, not forming neck .... *C. naqabensis*
        10. Flowers white or light lilac without stripes and veins
          12. Leaves 4–6, throat yellow.......... *C. caspius*
          12. Leaves usually 10–17, throat without yellow .... white forms of *C. pallasii* subsp. *pallasii*
    8. Style multifid
      13. Anthers white
        16. Tunics tough, splitting at base into narrow triangular teeth ...... *C. laevigatus*
        16. Tunics splitting at base in soft fibers
          17. Flowers lilac-blue, remaining open at night .... *C. tournefortii*
          17. Flowers white, closing at night.......... *C. boryi*
      13. Anthers yellow
        14. Flowers white, often striped or suffused violet on outside
          15. Leaves 3–4, tunics thinly papery with parallel fibers .... *C. veneris*
          15. Leaves 5–9, tunics splitting into many parallel fibers.......... *C. aleppicus*
        14. Flowers lilac.......... *C. serotinus* subsp. *salzmannii*

1. Corm tunics coarsely netted .... *C. serotinus* subsp. *serotinus*
1. Corm tunics finely reticulated
 18. Style divided into 6 or more slender branches .... *C. serotinus* subsp. *clusii*
 18. Style trifid
  19. Throat yellow
  20. Flowers mid deep purple inside, dark veined and/or buff-colored outside, bracteole absent .... *C. longiflorus*
  20. Flowers white or light blue, bracteole present
   21. Bract and bracteole greenish .... *C. niveus*
   21. Bract and bracteole silvery white
    22. Style branches exceeding half the length of the segments, throat glabrous, leaves up to 6 .... *C. asumaniae*
    22. Style branches less than half the length of the segments, throat pubescent, leaves 5–10
     23. Flowers white, leaves ciliated .... *C. hadriaticus* subsp. *hadriaticus*
     23. Flowers lilac, leaves glabrous or slightly papillose .... *C. thomasii*
  19. Throat without yellow
   24. Style branches nearly half as long as flower segments
    25. Style divides in throat .... *C. oreocreticus*
    25. Style divides well above the throat
     26. Flowers white
      27. Throat white, leaves up to 1 mm wide, style branches 10–16 mm long .... *C. hadriaticus* subsp. *parnassicus*
      27. Throat purple, if white, leaves 1–2 mm wide, style branches 7–10 mm long .... *C. mathewii*
     26. Flowers pale lilac, leaves 5–9.............. *C. hadriaticus* subsp. *parnonicus*
     26. Flowers pale lilac to reddish purple, leaves 10–17 .... *C. pallasii*
      28. Style branches pale yellow, anthers curved, enclosing style branches, perianth segments very narrow (up to 7 mm wide) .... subsp. *dispathaceus*
      28. Style branches red or orange
       29. Perianth segments rounded, style branches widely expanded, corm with fibrous neck up to 10 cm long .... subsp. *haussknechtii*
       29. Perianth segments acute, style branches expanding gradually, fibrous neck up to 6 cm long
        30. Perianth segments 4–12 mm wide, fibrous neck 3–6 cm long .... subsp. *turcicus*
        30. Perianth segments up to 20 mm or more wide, fibrous neck less than 2 cm long .... subsp. *pallasii*
   24. Style branches more than half as long as segments
    31. Segments 3.5–5 cm long, style branches 2.5–3.2 cm long, cultivated plant .... *C. sativus*
    31. Segments 1.4–3.3 cm long, style branches 1–2.7 cm long

32. Throat pubescent, style divided well below the base of anthers
    33. Segments at least 7 mm wide; leaves visible at flowering time .... *C. cartwrightianus*
    33. Segments 3–7 mm wide; leaves appearing after flowers ... *C. moabiticus*
32. Throat glabrous
    34. Flowers white or faintly lilac, style divided well above the base of anthers .... *C. asumaniae*
    34. Flowers mid lilac to purple, style divided in throat .... *C. oreocreticus*

This chapter begins with members of the *Crocus biflorus* group. There are comparatively very few autumn-blooming species (or subspecies) in Mathew's series *Biflori*. Most of them are leafless at flowering time and so were reviewed earlier. One of them—*C. wattiorum*—was discussed in the previous chapter, as my sample enters the winter with only minor leaves, but is also included here as its leaves are reported to develop together with the flowers. Currently, only one leafy autumn-blooming species remains in the *C. biflorus* complex and it is distributed in Greece in western and southern Peloponnesus. Another species, *C. caspius*, isn't that closely related and in the wild occurs near the Caspian Sea. Because far fewer people travel to Turkey and western Iran in late autumn than in spring, it is possible that several autumn-bloomers are yet to be discovered in this area.

Crocus melantherus Boissier & Orphanides ex Maw

PLATES 61 & 62

Crocus melantherus (2*n* = 12), meaning "black-anthered," was regarded as *C. biflorus* subsp. *melantherus* by Brian Mathew, but the plant and its name were known long ago. It is widely distributed and locally very abundant in western and southern Peloponnesus where it grows in short grassland and sparse maquis on limestone formations from 700 to 1200 meters.

The taxonomy of the *Crocus biflorus* group is still very unclear. Because of the scarcity of more detailed studies, Mathew degraded many of earlier species to the subspecies level. Many researchers, including Erich Pasche and Helmut Kerndorff, follow Mathew in describing new taxa in this group as subspecies, and possibly in many cases they are right. I prefer to keep the older names of many of the former species. My reason is quite practical—I need a smaller label for a pot and for the names of the picture files on my computer.

Crocus melantherus was grown in gardens for many years under the wrong name *C. crewei*, and under this name is included in *Flora Europaea* (Tutin et al. 1980). The true *C. crewei* is a spring-blooming crocus growing wild in western Turkey.

The first corms of *Crocus melantherus* that I grew came to me under the name of *C. crewei*, and only Mathew's monograph helped me to correctly identify the plants in those early years. From the beginning, this crocus became my favorite because of its beautiful black anthers, although it bloomed quite rarely—only in seasons with a long warm autumn—and I soon lost it. When I built a greenhouse it started to bloom regularly and then it turned out that not all plants had an equally prominent black connective on the anthers; in some plants the connective was quite narrow and pale and in a few specimens it was even absent. I observed the same thing in the wild.

The flowering time turned out to be very long. Occasionally in mid-February, when the temperature outside was well below 0°C, some pots in my greenhouse were filled with beautiful flowers of *C. melantherus* which had started to bloom in early December. In January when the temperature in the greenhouse dropped to −12°C, the plants were covered by glass-wool sheets for protection, and they still looked very nice when I opened the pots in early spring.

Flowers of *Crocus melantherus* almost always have a white background. Rarely have I seen specimens that were slightly lilac shaded. The outer side of the tepal is marked with heavy purple or gray stripes or sometimes nicely speckled with small dots. In some populations such speckled plants are very rare, but I have heard reports of localities where striped and speckled plants are in equal proportions. Sometimes the background color on the outside of the sepal is a creamy yellow, creating the impression of another "yellowish" autumnal crocus besides *C. scharojanii*, but this tone fades with age and by the end of flowering has returned to its "normal" white color. The leaves are grayish green without prominent ribs in the grooves on the lower surface. They appear with the flowers, and by the end of blooming can reach even the tepal tips. The throat is from grayish yellow through dark yellow to bright orange in the best samples.

This crocus is easy to grow in pots in a standard soil mix. It does not need a hot summer rest, so I usually move my pots out of the greenhouse during the summer heat. In Greece I saw many thriving populations of *Crocus melantherus* along the side of the road, but the farther I got from the road, the sparser the populations became. I suspect that there they were overgrazed by the ubiquitous goats and sheep.

Crocus caspius Fischer & C. A. Meyer ex Hohenacker

PLATES 63 & 64

Along the south shore of the Caspian Sea, in the Talish Mountains and at the foothills of Elburz is distributed another member of series *Biflori*, namely, *Crocus caspius* ($2n = 24$). According to Alan Edwards, it can be found from elevations of about 30 meters below sea level (the water level of the Caspian Sea is now much lower than that of the world's oceans) up to 1300 meters above sea level. It grows in woods and grassy scrubs from Gorgan in Iran to Lenkoran and Baku in Azerbaijan in humus-rich damp earth or sand or in grass.

While traveling in northern Iran, Admiral Paul Furse and his wife noted that the species was "common, dug by pigs." I searched for it during my trip to Talish (Azerbaijan part) in 1987 but failed to find it. A few years ago I obtained it from several sources, but all my bulbs are from Iran. Alan Edwards claims it as "one of the aristocrats of the genus." In 1903 it was awarded a Botanical Certificate from the Royal Horticultural Society.

Compared with other species of series *Biflori*, *Crocus caspius* looks like an outsider, so it was not surprising that phylogenetic studies showed it to be closer to the Central Asian species (series *Orientales*). Like all Central Asian crocuses its seed capsules ripen at ground level and are not pushed upwards by an elongating scape. They are easy to collect if ants do not take them away before you notice the dehisced capsule deep within the leaf rosette. The corm tunics do not have rings at the base but are membranous, splitting longitudinally, and they are not distinctly fibrous. Because of these two possible interpretations of the bulb tunic characteristics, I have included this species twice in the key.

In my specimens, the flower color varies from the purest white to light violet (sometimes offered under the name var. *lilacinus*) with a large invariably deep yellow-orange throat. A. A. Grossheim in *Flora of Caucasus* identified yellow as the color of the stigmatic branches of *Crocus caspius*, whereas Per Wendelbo in *Flora Iranica* noted that the stigma was invariably orange. In my white-flowered plants the stigmatic branches are yellow; in my bluish flowered plants, bright orange, and they well exceed the bright yellow anthers. The flowers are slightly fragrant. The leaves start to develop before the flowers and during the blooming they overtop them. My favorites are the white-blooming forms. Ian Young (Scotland) has observed that *C. caspius* has two waves of blooming, and when the first blooms start fading, the next set of flowers shows up. I have noticed that the later-developed flowers in my light lilac plants are a much deeper lilac shade, but I think that it is due to the much lower temperatures at the end of blooming.

The climate in the south Caspian region is relatively moist, so in cultivation a dry dormancy should be avoided. Alan Edwards stated that plants lost vigor when confined exclusively to pots and could be rejuvenated by a period of convalescence in a bulb frame where the roots were allowed to wander unfettered. This species sets abundant seed, which is green tinted when fresh and germinates very readily.

E. A. Bowles discovered that, despite its southern origin and the fact that it grows at low elevations, *Crocus caspius* turned out to be surprisingly hardy in the United Kingdom, blossoming freely in the open garden. Up until recently, it was a very easy plant for me to grow in pots, and it bloomed wonderfully every autumn. I kept the plants under cover even in summer, and replanted and divided corms every summer. But, during the moderately cold winter of 2008–09, a frost killed all three of my stocks of this species. Possibly I put on the winter cover too late as the strong frost started very suddenly. Winters in its native habitat are not very harsh due to low elevations and the closeness of the sea.

Crocus goulimyi Turrill

PLATES 65–69

Crocus goulimyi ($2n = 12$) is one of few species in the genus that can be identified by its corm. It has thick, hard, polished corm tunics and resembles a hazelnut more than a crocus corm.

The species was first collected in 1954 by C. N. Goulimy. Its area is limited within southern Peloponnesus at elevations from 30 to 750 meters, but wherever it grows it carpets the ground with marvelous blue flowers. Very few autumn-blooming crocuses can compete with it in beauty. It is quite surprising that it was not discovered earlier. Perhaps the reason is that in the past southern Peloponnesus has not been as popular a tourist district as it is today, especially in autumn when the bathing season ends.

The flowers are deeper or lighter lavender-blue with wide beautifully formed flower segments, the inner ones usually being lighter. They smell quite strongly of honey in warm weather.

A pure white specimen discovered on Mani Peninsula turned out to be an excellent grower and, after being propagated in the skillful hands of Michael Hoog, was named **'Mani White'** (Plate 66) The selection has acute petals which give the flower a well-defined triangular shape when compared with other forms of the species. Unfortunately, many long-cultivated and vegetatively propagated clones are virus infected today and therefore a careful selection is required.

John Fielding sent me an excellent plant with considerable contrast between the outer and inner segments; it is named **'Harlequin'**. During a trip to Peloponnesus on Mani Peninsula I found a nice group of similarly colored flowers formed by several corms. The flowers had an even more pronounced contrast in which the inner tepals were almost white but the outer ones were quite deep lilac. I nicknamed it **'New Harlequin'** (Plate 67). Somewhat to the west from Aeropolis my step-daughter Līga found a specimen with a distinctly pinkish flower, very different from all the others seen till now. The stigma is three-branched, yellow to orange (I have never seen a white stigma even in pure white flowers, but such have been reported by other travelers), and mostly shorter than the anthers (I have seen only a few specimens where it slightly exceeded the anthers).

Southwards from Monemvasía occur plants with lighter toned, almost white or light lilac flowers which are also somewhat smaller in size. They were separated by Brian Mathew as **subsp. *leucanthus*** (Plate 68). During a visit to Monemvasía and further south in the same population, I saw lighter and darker specimens of various sizes. According to Mathew (pers. comm.),

> Monemvasía is probably the place where the two subspecies merge, with subsp. *leucanthus* going south and subsp. *goulimyi* southwest to Mani Peninsula. Apart from the pale color of "pure" *leucanthus* there seems to be a slight difference in flower shape.

Later, subsp. *leucanthus* was lowered in its rank to a variety and therefore I have not included it in the identification key. This decision is supported by the fact that in the trade are offered plants under the name var. *intermedia* (Plate 69). I really cannot draw any distinguishing lines between all these taxa although there is an excellent picture of a large very uniform group of subsp. *leucanthus* growing southwest of Monemvasía (Mathew 2002).

On sloping banks of farm fields near Fotia, Peter and Penny Watt (2008) found specimens with flowers growing 30 cm above ground level and with leaves 60 cm long. Although this site is the locality of subsp. *leucanthus*, the Watts found plants in a wide range of colors, from occasional blue specimens as deep a blue as in subsp. *goulimyi*, to bicolored and very pale specimens. For the time being, I prefer to keep distinct color variants under cultivar names only.

I received my first corms of *Crocus goulimyi* in the 1980s from Michael Hoog and lost them all the first winter. I tried again a decade later when I built my first greenhouse. It turned out to be a vigorous grower, and in fact multiplied so quickly that I was forced to replant it annually. Mathew (1982) noted that "it is much better in a bulb frame, although its rate of increase can be almost too rapid when conditions are just right." My plants start blooming in midautumn (October), and the last flowers emerge simultaneously with hard frosts in December or January, but are then surpassed by the long leaves and are no longer spectacular.

Crocus laevigatus Bory & Chaubard

PLATES 70–72

The corm tunic of *Crocus laevigatus* ($2n = 26$) is almost as smooth and coriaceous as that of *C. goulimyi*, but not as tough and it splits at the base into long narrowly triangular teeth. The two species are easily distinguished, however, by the many-branched stigma of *C. laevigatus*.

In the wild this species grows in Greece, on the Cyclades and on Crete from sea level to 600 meters, but on Crete up to 1500 meters in open stony and rocky places, in sparse scrub, and occasionally in pine woodlands. When E. A. Bowles wrote his crocus handbook in 1924, *Crocus goulimyi* had not yet been discovered, and so he characterized *C. laevigatus* as a species that was unique in its "almost woody" corm tunic which was so hard that it decayed slowly. Near Athens he collected a corm with 15 tunics, the product of as many years. The specific name is derived from the Latin *laevis*, which means "smooth."

Crocus laevigatus represents one of the three closely related species joined by Brian Mathew in series *Laevigati*. They all have a similar distribution area in southern Greece, on islands as far as Crete, and phylogenetic studies confirm their close relationship. Sometimes their distinction is quite problematic. In my opinion the most creative solution to this problem comes from Peter and Penny Watt (2008):

> For us, though, an essential adjunct to growing crocus from series *Laevigati* is a two-sided label . . . . one side of the label claims *C. boryi*, the other *C. laevigatus*. We reverse the label whenever we change our minds. This system came into being in order to cope with *C. tournefortii* on Rhodes. . . . On rare occasion we look at the most confusing of crocus in series *Laevigati* and wonder whether we should not upgrade our system; but then where can we obtain a three-sided label?

Although I don't believe that it is difficult to distinguish between typical plants, problems could arise with the marginal variations in mixed populations. An exact identification might not be easy without seeing a corm, especially when a population of pure white *Crocus laevigatus* grows next to a population of *C. boryi*. In those circumstance, it helps to look at the shape of the flowers; in full sun, flowers of *C. laevigatus* open like flat stars, while flowers of *C. boryi* retain their goblet shape.

The flower color of *Crocus laevigatus* varies but usually is white. I have only rarely seen bluish flowers, and the darkest of those can be characterized as light blue. The exterior of the outer sepal color varies from pure white through variously purple striped to almost deep purple with a diffused wide light lilac edge. In most cases there is one longer median stripe with shorter stripes on either side and with finer lateral veining. In some cases the veins are so dark that they are visible through the sepals; looking at them when the flower is open gives the impression that there is darker veining on the inside, too. In some forms the exterior of the sepals is pleasingly light yellow, even golden or somewhat buff-toned.

The shape of flower segments can vary widely from narrow and pointed, forming starry flowers, to wide and rounded, producing beautifully rounded blooms. The flower size is very variable as well, so excellent forms could be selected for wider growing.

On the Cyclades are forms that bloom in early spring and, according to Brian Mathew, they retain this habit in cultivation. Forms from the mainland are strongly fragrant. In fact, E. A. Bowles listed *Crocus laevigatus* among the most fragrant species.

Commercial catalogs usually offer a form under the name **var. *fontenayi***. It has wide, rounded tepals of a beautiful light lilac color with darker striped sepal outsides and a bright orange throat. This form is comparatively late flowering. In cultivation also is a form with an unstriped light golden yellow sepal back.

The throat and filaments of *Crocus laevigatus* are yellow, but the anthers are invariably white. The style is divided into many light yellow to orange stigmatic branches. In the samples that I grew earlier, the style branches always well exceeded the anthers, but on Peloponnesus I often saw specimens with style branches ending below or at the tips of the anthers.

In the wild I have found this species in stony and rocky places, in clearings in sparse scrub, quite often by a roadside where sometimes it bloomed in a graveled strip between asphalt and grass. Reportedly it is a completely hardy species, but I grow it only under cover in raised beds or pots where it blooms starting from the first week of October and ending in December. I have never had forms blooming in spring, but at the Royal Botanic Gardens, Kew, I saw a specimen from Chios Island (Greece) blooming with distinctly lilac flowers at the same time as the latest-flowering spring crocuses. In my climate, it can suffer from long and harsh winters.

Crocus boryi J. Gay

PLATE 73

Superficially *Crocus boryi* ($2n = 30$) is most similar to *C. laevigatus* and often grows with it in mixed populations. It is named after French botanist Jean Baptiste Bory de Cent-Vincent. The species comes from western and southern Greece, some of the

Ionian Islands, and southeastern Crete, where it grows from sea level to 1500 meters. It can be found on undisturbed stony edges of fields, in uncultivated olive groves, open rocky hills in grass or scrub, and on grassy sand dunes.

The flowers are almost invariably white, though occasionally purple veins occur at the outer base of the flower segments, in which case the flower is usually larger and more wineglass-shaped than the flower of *Crocus laevigatus*. Rarely do flower segments with a purplish feathered back appear.

Both *Crocus boryi* and *C. laevigatus* are easily distinguished by the shape of the corm tunic. In *C. boryi*, as in its ally *C. tournefortii*, the corm tunics are more membranous, splitting at the base into many parallel fibers. The flower of *C. boryi* never opens as widely as it does in *C. laevigatus* or *C. tournefortii*, retaining the goblet shape even in bright sunshine, and the petals are more rounded. The goblet-shaped flowers of *C. boryi* also separate it from the white-flowered Greek species *C. hadriaticus*, but when in doubt, the white anthers and divided style of *C. boryi* further set it apart. Problems may arise with plants from Crete where some specimens of the closely allied *C. tournefortii* have white flowers and sometimes hybridize with *C. boryi*.

My first samples of *Crocus boryi* came from Peloponnesus, where I found plants growing in clearings among spiny scrubs on quite rocky formations. The flower size is quite variable with the tepals being as long as 1.5 cm in dwarf forms and up to 6 cm in giants. The throat is usually bright yellow, and in the best specimens may be orange with a diffused yellow zone surrounding the throat. The many-branched stigma is yellow to orange and slightly overtops the anthers, but near Aeropolis in southern Greece I found a population where the bright orange stigmatic branches significantly overtopped the anthers of all the plants.

Flowers of *Crocus boryi* vary considerably in flowering time. My earliest forms from Mistra started to bloom at the end of September, but those from Niniandra, only in December. In 2004 I saw a container-grown specimen of *C. boryi* blooming at an Alpine Garden Society show on 22 February. The color was an unusual light creamy yellow with an inconspicuous purplish midvein on the outside of the sepals. A similar-looking plant was selected among seedlings of *C. boryi* by Alan Edwards and named **'Brimstone'**.

Crocus boryi is reportedly less hardy than its allies, but this depends on the population. At present I have many samples of it; some grow problem-free in containers, others can suffer in unfavorable winters even under cover, as it happened in the winter of 2008–09. I keep the pots inside during the summer to provide warm and dry conditions.

Crocus tournefortii J. Gay

PLATE 74

If *Crocus laevigatus* and *C. boryi* are known to generally have white flowers, the third member of the family—*C. tournefortii* ($2n = 30$)—usually has lilac-blue flowers. When occasionally it does have white flower segments, it is easily distinguishable from its closet relatives: it is one of only three crocuses whose flowers once opened do not close anymore even at night.

The species was named after French botanist J. P. de Tournefort. Until 1980 it was known only from the islands of the Greek Archipelago. In 1980, following an earlier report by J. R. Marr, it was discovered on the north coast of Crete by Chris Brickell and Brian Mathew. It grows in stony ground and rock crevices or in dryish scrub usually on limestone formations, occasionally on mica-schist from sea level up to 650 meters.

In most cases the flowers are light lilac with a beautiful yellow throat, in the best samples edged by a diffused white zone before the color changes to lilac. On Crete, where white-flowered forms are more common, the pubescence of the filaments helps identify the species: filaments of *Crocus tournefortii* are strongly pubescent while those of *C. boryi* are only minutely papillose. The style in *C. tournefortii* is many-branched, bright orange or reddish, and greatly exceeds the length of the anthers. In most of the samples that I grow, the style reaches the tepal tips or surpasses them. In contrast, the style is much shorter in the other two related species.

On Crete *Crocus tournefortii* sometimes hybridizes with *C. laevigatus*. I found several blooming crocuses in a wet ditch along the road to the Lasithi plain. My arrival was a little too early as the autumn rains still had not started and the soil on the mountain slopes was very dry. Only the ditch had accumulated some water, and this had encouraged several plants to bloom, namely, *Colchicum* sp., *Sternbergia lutea (sensu S. sicula)*, *Narcissus serotinus*, and *Crocus*. In sparse scrub on the slope higher up the road, I found only a few *Cyclamen graecum* subsp. *candicum* blooming in rock crevices or in very spiny shrubs. Crocus corms and other bulbs were washed down by rains from the upper slopes. There white and light violet plants bloomed side by side. I identified them as *Crocus tournefortii* just by the length of stigmatic branches and the pubescence of the filaments. I was very confused the next season when the flowers bloomed but then closed in the evening at sunset. After some correspondence with Brian Mathew, we concluded that they most likely were hybrids between the two species.

Crocus tournefortii is one of the most attractive autumn-blooming crocuses but can be grown only in pots because of its habit to keep the flowers open even in rainy weather. Most possibly this feature has developed because of a specific pollinator which is active in the darkness. In summer *C. tournefortii* needs dry and hot conditions.

Crocus hyemalis Boissier & Blanche

PLATE 75

Two species growing wild in the Middle East and described from Syria have similar membranous corm tunics that split lengthwise into parallel strips: *Crocus hyemalis* and *C. aleppicus*.

Crocus hyemalis ($2n = 6$) is very rarely cultivated and blooms in November to December, rarely in January. Its name means "winter crocus." The species is common in Israel and in the adjacent regions of southern Lebanon and Jordan, where it grows on rocky hillsides mainly in scrub and in fallow fields, mostly at low elevations.

The flowers usually are white with several purple stripes on the outside base of the sepals; rarely are they entirely purple on the outside. The throat is comparatively large, bright orange fading towards a somewhat yellowish starry margin. The style is deeply divided into many branches and is approximately the same height as the anther tips. The most imposing feature is the black anthers, although sometimes the black color is slightly hidden when the yellow pollen grains burst out. Helmut Kerndorff in Jordan observed some variation in the anther color and found in the inspected population of 78 individuals that 72 had mostly yellow anthers with a thin black margin, 5 had plain yellow anthers, and only one was almost entirely black.

I have never had this species in my collection, but John Lonsdale (United States) is one of a couple of enthusiasts who successfully grow it. Lonsdale's plants have such a bright orange throat that it shines through the tepals and can be seen even in closed flowers. The base of the outer flower segments is marked by short purple stripes; sometimes only one very narrow middle stripe reaches the tepal tips. The species can be grown only in a bulb frame or greenhouse and needs hot and dry summer conditions.

Crocus aleppicus Baker

PLATES 76–78

Crocus aleppicus ($2n = 16$) was described from grassy sites near Aleppo in Syria. It is a plant of exposed stony places in maquis, sparse grass, or scrub, often on basalt formations but also on sandstone and limestone, from 50 to 1750 meters.

Although *Crocus aleppicus* and *C. hyemalis* are similar in flower color, shape, and blooming time, they can be easily told apart by their leaves: those of *C. hyemalis* are much wider with a broad white very striking stripe on the upper surface while those of *C. aleppicus* are much narrower with a very pale, almost indistinct median stripe. The anthers of *C. hyemalis* usually are black; in *C. aleppicus* they are yellow or blackish before they dehisce. The throat of *C. aleppicus* is paler yellow to light lemon. Additionally, the corm tunics in *C. hyemalis* are very thin, membranous, and more prominently splitting at the base, but in *C. aleppicus* are strongly fibrous, separated for much of their length, and only inner tunics are membranous.

Crocus aleppicus has little horticultural value because of its narrow pointed flower segments and starry appearance, but it is extremely variable in both the color and tepal shape. Flowers of my plants had a better-than-typical shape and had more rounded tepal tips, as do some acquisitions in John Lonsdale's collection. In one of Lonsdale's specimens the sepal outside is completely speckled or diffusely veined light grayish purple. Brian Mathew reported a plant in which the outer segments have a central band of purple edged yellow. I had this species twice, but in both cases lost the corms within a few years. I suppose that the winter conditions here are too dark and wet for it. In summer this crocus requires very dry and hot conditions and it can be grown only under cover.

Much more successful with *Crocus aleppicus* are H. Kerndorff and E. Pasche. For them the greatest problem is *Botrytis*, which makes continuous and ample ventilation very essential. The plant needs very well drained soil to prevent corm rotting in winter. Plants from high mountain regions of Jordan seem to be more vigorous than plants from the lowland. Although this crocus has set seeds for both German crocus specialists, it still remains rare in cultivation.

Crocus veneris Tappeiner ex Poech

PLATE 79

Crocus veneris ($2n = 16$) is named after the Roman goddess of love, Venus, and is widespread in Cyprus, where it grows in open stony or grassy places and among scrub and in sparse woodland from 120 to 1000 meters. The species is most closely related to *C. aleppicus*, but easily distinguishable by the number of leaves: *C. veneris* usually has three or four, *C. aleppicus*, five to nine. The corm tunic of *C. veneris* is much finer, less fibrous than that of *C. aleppicus*. Flowers, however, are very similar. Those of *C. veneris* usually are white and vary only in the amount of purplish veining at the outer base of the tepals.

In one of my samples the flowers have only a few small purple stripes; in the other, a wide purple "tongue" with the tip reaching one third of the tepal height. The throat is somewhat smaller, yellow with a less starry edge than in *Crocus aleppicus*. The anthers usually are yellow, occasionally purplish brown. The leaves generally appear simultaneously with the flowers, but during flowering they quickly lengthen and in the second half of blooming surpass the flower. In my opinion this is the biggest fault of this otherwise charming little species.

Crocus veneris is quite easy to grow and much easier than its similarly looking relatives from Syria and Israel. Due to the late flowering and the weather conditions in its homeland, it is suitable only for pots or a frame where a hot and dry summer can be ensured.

Crocus ochroleucus Boissier & Gaillardot

PLATES 80 & 81

Crocus ochroleucus ($2n = 10$) is closest to *C. kotschyanus*, which was treated in the previous chapter, but the two species are very easily differentiated since *C. kotschyanus* in most forms never produces leaves in autumn.

In all forms of *Crocus kotschyanus* the yellow zone in the throat is always formed by two spots at the base of the tepals, although they can be confluent forming a deep yellow ring around a paler throat. In *C. ochroleucus* the throat is uniformly yellow, in some specimens it is even bright orange shining through the tepal bases on the outside. The species name is derived from the flower color. The corms of *C. kotschyanus* are flattish and somewhat irregular in shape, while those of *C. ochroleucus* are smaller and rounder. Corm tunics are thinly membranous with fibers parallel to the base and slightly reticulated at the top, not as smooth as in *C. kotschyanus*. Like most forms of

C. kotschyanus, *C. ochroleucus* produces an abundance of small cormlets on the surface of a new corm every season. The phylogenetic tree confirms its close relationship with the type subspecies of *C. kotschyanus*.

Crocus ochroleucus occurs in the wild from southwestern Syria through Lebanon to northern Israel, growing on limestone or basalt formations in rocky or stony areas, sometimes in oak scrub from 300 to 1500 meters. In my collection the flowers start to bloom in the second half of October or in November and are abundant although not very large. The white anthers easily separate this species from white forms of the Spanish crocuses, but the trifid stigma well distinguishes *C. ochroleucus* from its neighbors. The flower segments in typical plants actually are creamy white and the best of them have a deep yellow-orange throat.

In some forms the outside of the sepal is slightly grayish shaded. The Tel Aviv Botanic Garden selected an excellent albino form with only a slightly darker shaded throat; later the selection was named **'Valerie Finnis'** after a famous gardener. According to Oron Peri this variety makes small clumps and pure albino forms are not uncommon. There is a hybrid with *Crocus kotschyanus* (see *C. kotschyanus* for details of the hybrid).

In milder climates *Crocus ochroleucus* does well in an open garden. Brian Mathew grew it successfully even in grass. In my collection, it must be grown only in pots. According to Mathew, it has a tendency to split into many small corms; to avoid this he recommends deeper planting in the same way as is done with *Iris danfordiae*. My experience with this species is fairly limited because all the samples I received from commercial nurseries in last years turned out to be infected with virus. Only recently did I obtain a healthy stock from the personal collection of Robert Potterton, owner of Pottertons Nursery.

Crocus cambessedesii J. Gay

PLATES 82 & 83

Crocus cambessedesii ($2n = 16$) is so special that it is almost impossible to misidentify. It was named after French botanist Jacques Cambessedes. The species is very widespread on Majorca and Minorca islands where it grows in pine woodlands and in scrub or on open rocky hillsides up to 1500 meters above sea level.

Plants in my collection begin to flower in the second half of October and continue to March, so it was not easy to decide whether to include the species among the autumn-bloomers or the spring-flowering crocuses. I have included it in the keys for both groups but, because blooming always starts in autumn and it enters the winter with well-developed leaves, I am treating it here with the autumnal crocuses. When I cover my plants for winter, this species is still in full flower and doesn't stop flowering until soon after I remove the winter cover.

E. A. Bowles treated *Crocus cambessedesii* with the spring-bloomers from the *C. imperati* group noting its close relation to those species. Phylogenetic studies confirm this opinion, putting *C. versicolor* as its closest neighbor and placing *C. cambessedesii* alongside *C. minimus*, *C. corsicus*, and *C. imperati*.

Flowers of *Crocus cambessedesii* are the smallest in the genus. The base color of the segments varies from white with usually creamy or buff-toned backs on the outer tepals to very deep lilac. The throat and filaments are invariably white, the anthers deep yellow. In darker forms the white part of the throat is very small and the filaments are light violet shaded. Sometimes the throat is surrounded by short purple stripes. The bright orange or red stigma is deeply divided into three branches, somewhat frilled or slightly lobed at the top, usually below the anthers and only rarely reaching their tips. All forms that I have seen had deep purple stripes on the back of the sepals, but Brian Mathew listed occasional albinos without stripes.

In pots this species blooms very abundantly and is one of my favorites. Brian Mathew stated that it is too small for an open garden but delightful in an alpine house or bulb frame. In my climate it can be grown only under cover and must be provided with dry and warm summer conditions.

Crocus niveus Bowles

PLATES 84 & 85

If *Crocus cambessedesii* has one of the smallest flowers among the autumn-blooming species, *C. niveus* ($2n = 24$) has one of the largest flowers, not only among the autumn-blooming crocuses but also among spring-bloomers as well. Its flowers were initially described as invariably white and for this reason it got the name *niveus* which means "snow-white." Subsequently it was discovered that flower color is quite variable. In fact, in commerce a "blue form" is offered more often; its flowers are nearly white, but only nearly, with a clearly distinct very light blue shade. In autumn 2008 I found a couple of samples with much deeper blue flower segments, but such are extremely rare though not less beautiful than the pure white ones. This species is very common in southern Peloponnesus where it grows in olive groves and in scrub on limestone formations at 50 to 750 meters.

The largest population that I have seen was near the road from Neapolis to Velanidea on very rocky ground covered by a low, very spiny shrub. Never before have I seen crocuses in such a dense groundcovering. There were almost no pathways between shrubs, and the crocus flowers were interlaced with the low, spiny branches of the scrub. Fortunately, I was wearing very good mountain boots with a solid, thick sole and could simply step on the thorny shrubs. Taking a photo wasn't easy, however, as I had to kneel and my jeans didn't have the same kind of protection that my boots did. Most flowers had already bloomed, yet still I found hundreds and hundreds in bloom close by the road. They were in all shades, mostly very light lilac similar to the "blue form" in commerce, but also many pure white and very few violet, only the inside of the flower segments was lighter. The flower throat was invariably deep yellow but the stigma was rather variable from light yellow to almost red. The branching also varied—some were just clearly trifid while others were divided at the tips into many short branches. Growing nearby in more open places were plenty of *Crocus goulimyi* and, by the roadside, large groups of *C. laevigatus*. Peter and Penny Watt

(2008) report some specimens without any yellow in the throat and with dirty white style branches.

In other populations seen during my trip plant variability was not so great—flowers were mostly white with some light blues among the whites. I saw *Crocus niveus* only in densely overgrown places or even under trees in the forest. Most of olive groves and forests along my route were been burnt down by the previous years' disastrous forest fires and there was not a single crocus to be found among the tree remnants. I was shocked to drive for hours on end and see nothing but a black landscape with some trunks like fingers raised to the sky.

In my collection, *Crocus niveus* is one of the easiest autumn crocuses growing under cover. Brian Mathew reports that it is a very vigorous garden plant in sunny well-drained positions. It increases quickly and sets abundant seed, even in the greenhouse, if pollinated by bees on sunny days when the flowers have nicely opened.

Crocus longiflorus Rafinesque

PLATE 86

Despite its name, which means "long-flowered," *Crocus longiflorus* ($2n = 28$) does not have the longest flowers in the genus. In fact, many species surpass it in this respect. A more appropriate name would be *C. odorus* as the flowers are very fragrant, but that name is not possible according to the rules of priority that govern plant naming. According to E. A. Bowles, only *C. laevigatus* and *C. imperati* compete in fragrance with *C. longiflorus*. In the wild *C. longiflorus* is distributed in southwestern Italy, Sicily, and Malta where it grows mainly on limestone formations in dry grassland and forest fringes from 200 to 1900 meters.

Flower color can be very variable, though I have come across specimens with mostly more or less uniformly dark lilac-colored blooms. The sepal outside can sometimes be striped, and it is possible to find specimens in which the deep purple stripes surround a bright orange throat. Sometimes like in almost any crocus species albinos can be found but even these have a bright yellow throat.

E. A. Bowles (1952) could only express his dream about a pure white *Crocus longiflorus*. Now, two such forms are rarely available. The first was distributed by Michael Hoog under the name **'Albus'** and it was of purest white. The other, known under the name **'Primrose Warburg'**, has purple stripes on the sepal outside base and flower tube and it was originally found on Malta. Both are real beauties. The style is deeply divided into three light or dark orange branches, very rarely yellow. The filaments and anthers are yellow.

Crocus longiflorus is one of the hardiest species of this group and for several years it overwintered in my collection in the open garden (before my "greenhouse era" started), although it flowered only once. Plants had barely shown green leaf tips before winter frost set in. Now they do much better under cover where the tender leaves are protected. Plants must be kept dry in summer.

Crocus serotinus Salisbury

Crocus serotinus and its three subspecies grow wild in the western part of the crocus range—in Spain, Portugal, and northern Africa. The plants are very variable, leading E. A. Bowles (1952) to treat them as four different species segregated along with the leafless *C. nudiflorus* in the Spanish group. They are quite widely grown and some are regularly offered in commercial catalogs, although again I have to urge you to carefully check the flowers and leaves after receiving a new stock as quite often the stocks (especially those offered as subsp. *salzmannii*) are virus infected. The specific name means "late coming" and corresponds with the late flowering habit in the wild prior to the winter season.

Crocus serotinus* subsp. *serotinus

PLATE 87

The type subspecies, *Crocus serotinus* subsp. *serotinus* ($2n = 22, 24$), is the most distinct in its coarsely reticulated corm tunic. It grows in Portugal, from Lisbon southward to Algarve and along the coast to at least Cadiz in Spain, where it is to be found in sandy, acid soils on rocky formations and in pine woodlands near the sea up to 500(?) meters. Frequently, variants of subsp. *salzmannii* are grown under this name, but they are easily distinguishable just by the corm tunic which in subsp. *salzmannii* splits into parallel fibers.

Flowers are nicely scented, paler or deeper lilac, sometimes with darker veins; the throat is white or pale yellow. The style is divided into many short bright orange branches and usually exceeds the bright yellow anthers. In my plants the leaf tips appear somewhat after the start of blooming, but before the second flower emerges, and the throat is white, even slightly grayish shaded. This crocus needs a dry and warm summer rest and thus it has to be grown under cover.

Crocus serotinus* subsp. *clusii (J. Gay) B. Mathew

PLATE 88

Crocus serotinus subsp. *clusii* ($2n = 22, 23$) replaces subsp. *serotinus* in northern Portugal from the Lisbon area to northwestern and southwestern Spain as far as Cadiz. It grows in clay or humus-rich soils in partial shade under trees and in scrub or in fully open spots, usually on limestone formations at about 100 to 900 meters. Areas in which the two subspecies grow overlap, but the two can be distinguished by the much more finely netted tunic in subsp. *clusii*. It is also leafier and forms four to seven leaves per corm whereas the type subspecies has only three or four leaves. According to Brian Mathew the leaves in subsp. *clusii* appear somewhat later—when the flowering ends. I have observed just the opposite: the leaves in my samples of subsp. *clusii* during flowering are more advanced than those in subsp. *serotinus*. Flowers on my plants are quite similar in color, only the sepal outsides have a darker purple speckling and veining.

Like the type subspecies, subsp. *clusii* is not too often cultivated though both require similar conditions and in my collection can be grown only under cover.

Crocus serotinus subsp. *salzmannii* (J. Gay) B. Mathew

PLATE 89

The third subspecies, *Crocus serotinus* subsp. *salzmannii* ($2n = 22, 24, 44$), is the most widely distributed in cultivation and the most variable in the entire *C. serotinus* complex. Its corm tunic is membranous splitting into parallel fibers, and so it is quite easy to distinguish this subspecies from the others. In the trade are forms with a stoloniferous habit (very similar to that of *C. nudiflorus* stolons), and according to Brian Mathew this behavior is sporadic throughout the entire range of subsp. *salzmannii*.

The subspecies grows from North Africa to southern, central, and northern Spain, and it is no wonder that over the years it was described under several names. The most common of these are *C. asturicus*, *C. granatensis*, and *C. salzmannii*. In the wild, the habitats of this subspecies are very variable—from limestone rocks to more acidic sandy soils, sometimes on cliffs, on open grassy slopes, and in sparse pine woodland and scrub from 100 to 2350 meters.

The flowers are scentless but very variable in shape and color. In most forms they are paler or darker violet with a creamy white to deep yellow throat. The outside of the flower segments can be plain-colored or more or less intensively veined. Likewise, leaf development varies greatly. In some forms the leaves are well developed at the time of flowering; in others only the leaf tips have started to show when the flowers occur. The form known under the name **'Erectophyllus'** is leafier and originates most probably from northern Africa. Unfortunately I haven't succeeded in finding a virus-free stock of it. The same is true for a cultivar distributed under the name **'Atropurpureus'**.

Albino forms can be found in each subspecies and, according to Brian Mathew, white-colored specimens with a purple throat have been discovered. Two albinos are now in cultivation: **'Albus'** with a light yellow throat and **'El Torcal'** with a white throat. The latter was originally collected at an elevation of 1000 meters by Henning Christiansen in southernmost Spain near Antequera. It is one of the best white-colored cultivars of subsp. *salzmannii*, though not a very fast increaser. Several years ago it was offered in some catalogs but at present I cannot find it anymore.

The plants from Morocco are larger, but those from the northern part of the range are more vigorous and easier in gardens. Chris Brickell grows this subspecies in a lawn in his West Sussex garden. Many years ago it grew on the rockery of the Tallinn Botanic Garden (Estonia). For a number of years some forms in my collection survived in the open garden but eventually I lost them all. In pots it creates no problems and readily sets seed.

Plate 1. Raised beds with two rows each of *Crocus heuffelianus* in the author's current outdoor garden.

Plate 2. Raised beds of *Crocus biflorus* subsp. *weldenii* 'Miss Vain' in a field at Jan Pennings's nursery in the Netherlands. Photo: Jan Pennings.

Plate 3. *Crocus tommasinianus* naturalized in grass in the United Kingdom.

Plate 4. *Crocus veluchensis* and various cultivars of *Corydalis solida* in a formal bed at the Gothenburg Botanical Garden in Sweden.

Plate 5. The author's bulb collection includes crocuses planted in plastic pots set in raised beds in a poly-tunnel.

Plate 6. Drip irrigation of crocuses and other container-grown bulbs.

Plate 7. Plants in the *Crocus* collection of Michael Kammerlander are in clay pots plunged in coarse sand.
Photo: M. Kammerlander.

Plate 8. Removing winter cover from pots in the author's greenhouse.

Plate 9. Author pollinating crocuses. Photo: Klaus Stöber.

Plate 10. An open seedpod of *Crocus suworowianus*.

Plate 11. Bed with marked *Crocus veluchensis* seedlings.

Plate 12. Crocus flower deformation and color break in virus-infected plants.

Plate 13. Mosaic symptoms on leaves of *Crocus olivieri* subsp. *istanbulensis*.

Plate 14. Tobacco rattle virus symptoms often are not visible on all shoots emerging from the same corm.

Plate 15. *Crocus tauricus* corm tunics showing longevity of an individual plant in nature.

Plate 16. *Crocus kotschyanus* subsp. *cappadocicus* corms must be positioned vertically at planting.

Plate 17. Unlike most *Crocus* corms, those of *C. suworowianus* lay vertically in the ground.

Plate 18. *Crocus banaticus.*

Plate 19. *Crocus banaticus* 'Snowdrift'.

Plate 20. *Crocus speciosus* subsp. *speciosus* on Tschatir-Dag yaila in Crimea, Ukraine.

Plate 21. *Crocus speciosus* 'Cloudy Sky' selected from wild plants by Zhirair Basmajyan in Armenia.

Plate 22. *Crocus speciosus* 'Blue Web' raised by Leonid Bondarenko in Lithuania.

Plate 23. *Crocus speciosus* subsp. *ilgazensis.*

Plate 24. *Crocus speciosus* subsp. *xantholaimos.*

Plate 25. *Crocus speciosus* subsp. *archibaldii* (WHIR-129), from side.

Plate 26. *Crocus speciosus* subsp. *archibaldii* (WHIR-129), from top.

Plate 27. *Crocus pulchellus.*

Plate 28. *Crocus wattiorum* in the wild. Photo: David Millward.

Plate 29. *Crocus wattiorum* in the author's collection.

Plate 30. *Crocus nerimaniae*.

Plate 31. *Crocus kotschyanus* in the wild in Turkey. Photo: Kees Jan van Zwienen.

Plate 32. *Crocus kotschyanus* (HKEP-9205).

Plate 34. *Crocus kotschyanus* subsp. *cappadocicus* (BATM-372).

Plate 33. *Crocus kotschyanus* var. *leucopharynx*.

Plate 35. *Crocus kotschyanus* subsp. *hakkariensis* (HN-0114).

Plate 36. *Crocus suworowianus* (LST-304).

Plate 37. *Crocus vallicola* (HN-05) from Zigana Pass in Turkey.

Plate 38. *Crocus scharojanii* subsp. *scharojanii* from Dombai, Russian Caucasus. Photo: Dmitriy Zubov.

Plate 39. *Crocus scharojanii* albino form from 1900 to 2100 m in Dombai, Russia. Photo: Dmitriy Zubov.

Plate 40. *Crocus scharojanii* subsp. *lazicus* from Zigana Pass in Turkey.

Plate 41. *Crocus scharojanii* var. *flavus* (hybrid with *C. vallicola*).

Plate 42. *Crocus autranii.*

Plate 43. *Crocus autranii* (purple flowers) and *C. vallicola* (white flowers).

Plate 44. *Crocus karduchorum* from Bitlis in Turkey. Photo: M. Kammerlander.

Plate 45. *Crocus gilanicus*. Photo: John Lonsdale.

Plate 46. *Crocus nudiflorus*.

Plate 47. *Crocus robertianus*. Photo: M. Kammerlander.

Plate 48. *Crocus robertianus* var. *albus*.

Plate 49. *Crocus cancellatus* subsp. *pamphylicus* in the wild. Photo: Kees Jan van Zwienen.

Plate 50. *Crocus cancellatus* subsp. *mazziaricus* (LST-402).

Plate 51. *Crocus cancellatus* subsp. *lycius* (JATU-054).

Plate 52. *Crocus cancellatus* subsp. *damascenus* in the Negev Desert. Photo: Oron Peri.

Plate 53. *Crocus cancellatus* subsp. *cancellatus* on the Golan Heights. Photo: Oron Peri.

Plate 54. *Crocus cancellatus* subsp. *cancellatus* var. *cilicicus*.

Plate 55. *Crocus ligusticus* from Mount Biguone in Italy.

Plate 56. *Crocus hermoneus* subsp. *hermoneus* on Mount Hermon at 1700 m. Photo: Oron Peri.

Plate 57. Close-up of *Crocus hermoneus* (SBL-119). Photo: John Lonsdale.

Plate 58. *Crocus moabiticus*. Photo: John Lonsdale.

Plate 59. Dark form of *Crocus moabiticus*. Photo: Oron Peri.

Plate 60. *Crocus moabiticus* sometimes produces leaves and flowers at the same time. Photo: Oron Peri.

Plate 61. *Crocus melantherus* in Peloponnesus near Mount Killíni. Photo: Kees Jan van Zwienen.

Plate 62. *Crocus melantherus* (PELO-028) in Peloponnesus.

Plate 63. *Crocus caspius.*

Plate 64. White form of *Crocus caspius.*

Plate 65. *Crocus goulimyi* near Monemvasía in Greece.

Plate 66. *Crocus goulimyi* 'Mani White'.

Plate 67. *Crocus goulimyi* 'New Harlequin'.

Plate 68. *Crocus goulimyi* var. *leucanthus*, near Monemvasía village. Photo: Kees Jan van Zwienen.

Plate 69. *Crocus goulimyi*, Sikea form, intermediate between var. *goulimyi* and var. *leucanthus*. Photo: Kees Jan van Zwienen.

Plate 70. *Crocus laevigatus* (PELO-012) in Peloponnesus .

Plate 71. *Crocus laevigatus* at Cape Malea, Greece. Photo: Kees Jan van Zwienen.

Plate 72. Dark form of *Crocus laevigatus*.

Plate 73. *Crocus boryi* (PELO-020) near Mistra in Peloponnesus.

Plate 74. *Crocus tournefortii*. Photo: John Lonsdale.

Plate 75. *Crocus hyemalis* with prominent black anthers. Photo: Oron Peri.

Plate 76. *Crocus aleppicus* from the coast of northern Israel. Photo: Oron Peri.

Plate 77. *Crocus aleppicus* from the Mount Hermon population. Photo: Oron Peri.

Plate 78. *Crocus aleppicus* in the Golan Heights. Photo: Oron Peri.

Plate 79. *Crocus veneris.*

Plate 80. *Crocus ochroleucus* in the Golan Heights. Photo: Oron Peri.

Plate 81. *Crocus ochroleucus.* Photo: John Lonsdale.

Plate 82. *Crocus cambessedesii.*

Plate 83. *Crocus cambessedesii.*

Plate 84. *Crocus niveus* (PELO-016).

Plate 85. Blue form of *Crocus niveus* (PELO-014).

Plate 86. *Crocus longiflorus* from Malta. Photo: M. Kammerlander.

Plate 87. *Crocus serotinus* subsp. *serotinus* from Sagres in Portugal. Photo: M. Kammerlander.

Plate 88. *Crocus serotinus* subsp. *clusii.*

Plate 89. *Crocus serotinus* subsp. *salzmannii* (JP-9177) in the author's collection.

Plate 90. *Crocus sativus.*

Plate 91. *Crocus oreocreticus.*

Plate 92. *Crocus cartwrightianus* from Crete in the author's collection.

Plate 93. Some forms of *Crocus cartwrightianus* (CEH-613) look very similar to *C. sativus.*

Plate 94. White form of *Crocus cartwrightianus.*

Plate 95. *Crocus cartwrightianus* 'Purple Heart'.

Plate 96. *Crocus hadriaticus*.

Plate 97. *Crocus hadriaticus* subsp. *parnassicus* in the author's collection.

Plate 98. *Crocus hadriaticus* subsp. *parnonicus* color variant in Greece.
Photo: Kees Jan van Zwienen.

Plate 99. *Crocus thomasii*.

Plate 100. *Crocus pallasii* subsp. *pallasii* on Kaya-Bash in Crimea, Ukraine. Photo: Dmitriy Zubov.

Plate 101. *Crocus pallasii* subsp. *pallasii* from Crimea in Ukraine. Photo: Dmitriy Zubov.

Plate 102. White form of *Crocus pallasii* subsp. *pallasii* from Crimea, Ukraine. Photo: Dmitriy Zubov.

Plate 103. *Crocus pallasii* subsp. *pallasii* at Upper Galilee, Israel. Photo: Oron Peri.

Plate 104. *Crocus pallasii* subsp. *pallasii* 'Homeri' from Chios Island, Greece.

Plate 106. *Crocus pallasii* subsp. *dispathaceus* in Turkey. Photo: Kees Jan van Zwienen.

Plate 105. *Crocus pallasii* subsp. *haussknechtii* (WHIR-202).

Plate 107. Another form of *Crocus pallasii* subsp. *dispathaceus* in Turkey. Photo: Kees Jan van Zwienen.

Plate 108. *Crocus pallasii* subsp. *dispathaceus* (JKP 98-123) in the author's collection.

Plate 109. *Crocus pallasii* subsp. *turcicus* in Turkey. Photo: David Millward.

Plate 110. *Crocus pallasii* subsp. *turcicus* from Lebanon in the author's collection.

Plate 111. *Crocus asumaniae* in Turkey. Photo: Kees Jan van Zwienen.

Plate 112. *Crocus asumaniae* (JP-8783) in the author's collection.

Plate 113. White form of *Crocus asumaniae*.

Plate 114. *Crocus mathewii* in the wild. Photo: Kees Jan van Zwienen.

Plate 115. *Crocus mathewii* 'Brian Mathew'.

Crocus sativus Group

The remaining eight or nine species belong to series *Crocus* in Brian Mathew's monograph, and in E. A. Bowles's handbook they were reviewed in the chapter titled "The Saffron Crocus." It is a very complicated group in which the differences between several taxa are vague and overlapping and there is such great variation within the species that sometimes determining the correct name of a plant is quite problematic. The status of some now-recognized species is questioned by other botanists. I am not very satisfied with the identification key for this group of crocuses (included in the key on pages 80–82) as there are too many color variants, but I have tried to do my best.

Crocus sativus Linnaeus

PLATE 90

Our review of this group begins with one of the oldest cultivated plants, *Crocus sativus* ($2n = 24$), the source of the famous ancient flavor and drug we know as saffron. In olden times saffron was mostly used in medicine. It is mentioned by many ancient authors, in the oldest medical handbooks, in Egyptian papyruses, and in Indian medical manuscripts around 500 B.C. Saffron was widely cultivated in ancient Persia and Kashmir. Today you can buy top-quality saffron at almost every Iranian hotel. It is still widely used, though currently as a spice and dye and to a lesser extent in medicine.

The origin of cultivated saffron is still a mystery. The plant is a sterile triploid which multiplies only vegetatively. Most likely it is a selected form of *Crocus cartwrightianus*, although in the wild there are no plants with such huge flowers and such long stigmas as in the cultivated *C. sativus*. Similar forms can be found from Iran to as far as Italy. Extensive research was done in the former Soviet Union to find a replacement for *C. sativus*; it was discovered that, in terms of chemical qualities, *C. pallasii* from Crimea and southern Ukraine was the closest ally. Even *C. speciosus* was tested and acknowledged as a possible replacement.

Saffron has always been expensive, therefore it is not surprising that the drug has at times been adulterated by the addition of dried marigold tepals and other similar substances. Saffron forgers were punished severely. *Encyclopedia Britannica* refers to Nuremberg in the fifteenth century as a place where "men [were] being burned in the market-place along with their adulterated saffron, while on other occasion three persons convicted of the same crime were buried alive." In the Middle Ages saffron was something like a cure-all. In 1670 Johann Hertodt published a huge 300-page volume on saffron titled *Crocologia*. It had prescriptions for every illness using saffron along with such drugs as "dragon blood," "mountain mouth excrements," "swallow nests," and so forth.

Today it is still a great art to produce top-quality saffron. It takes one million blooms to yield approximately 10 kilograms of saffron, and the picking of all these styles is exclusively a hand job.

Crocus sativus is offered in many bulb catalogs, but to get a good flowering it must be grown in areas with a really hot and long dry summer. It does not flower annually in my collection, not even in the greenhouse. I think I have seen more autumns where my plants had only leaves than autumns with blooming flowers, though in places with longer summers I have seen it blooming nicely in an open garden. During the Soviet period I tried to grow corms from the Apsheron peninsula (Azerbaijan) in my garden. They bloomed the first autumn but with every lifting the corms became smaller and smaller until within three years I lost them.

Flowers of *Crocus sativus* are very large, bright violet-purple with darker veins and a deep purple throat. A clone with a more purplish shade is offered in the trade as **var. *cashmerianus***. The style is very deeply divided into three bright red branches which reach the tips of the flower segments or are longer. Unfortunately, as it often happens with very long cultivated plants that are propagated only vegetatively, most of the stocks that I've seen clearly show symptoms of virus infection.

Crocus oreocreticus B. L. Burtt

PLATE 91

If I am ever forced to downsize my collection, one of the species I could part with without much questioning is *Crocus oreocreticus* ($2n = 16$); it has the smallest flowers among its relatives. The species comes from high elevations in Crete, where it grows on rocky mountainsides in low scrub on heavy clay soils over limestone formations from 1000 to 2000 meters. Its name means "Cretan mountaineer."

The flowers are lighter or darker purple with prominent veining and somewhat pointed in overall appearance; the inner segments in my samples are remarkably shorter and narrower than the outer ones. The filaments are very short and white, the anthers much longer and bright yellow. The style divides in the throat into three bright orange to red branches which end near the tips of anthers. Leaves appear simultaneously with flowers or at the end of blooming.

In my collection this is the only crocus in which the cataphylls rise above ground to quite a great length, closely enwrapping the green leaves and flower stalk and only then on the top opens a flower. It seems like the flowers are being pushed out onto the tip of a pencil pressed into the ground. This feature is so unusual that I have never misidentified this species. Maybe I plant its corms too shallow? I have sometimes, but not always, seen similar plants in other collections. The flowers are of such special color and shape that it would be difficult to confuse this species with any other.

Crocus oreocreticus is a good grower in greenhouse conditions, multiplying by corm splitting and by setting seed. In my area, hand pollination is essential as the flower blooms so late that no natural pollinators are active at that time.

Crocus cartwrightianus Herbert

PLATES 92–95

Taxonomically *Crocus cartwrightianus* (2*n* = 16) is closest to *C. oreocreticus* and, like it, grows on Crete but at much lower elevations. It is widespread in Attica and the Cycladic Islands and on Crete and is named after Mr. Cartwright who worked in the British Consulate in Istanbul and sent some corms of this species to William Herbert. In the wild it grows in open spots and sparse woods or scrub on granite and limestone formations from sea level up to 1000 meters.

It is one of the nicest species in the group for its long bright red style branches, although in a few cases I have seen blooms where the style was shorter than half the length of the tepals. These were probably a seasonal variation and not very characteristic of *Crocus cartwrightianus*.

As previously mentioned, *Crocus sativus* allegedly originated from *C. cartwrightianus*. Many forms of the latter resemble the cultivated triploid plant in color, and in some regions the stigmas are collected for saffron. The color of *C. cartwrightianus* is extremely variable. In fact, the same population can include flowers that are entirely pale to deep lilac, others with more or less striped segments, and still others that are a solid pure white. The species is easily identifiable as it is one of only three with flowers that do not close at night. Of the others, *C. tournefortii* has a multi-branched stigma and *C. moabiticus* much narrower flower segments.

In the lilac forms of *Crocus cartwrightianus*, the throat color varies from almost white to deep purple through various degrees of purple striping. Especially nice are pure white forms in which the bright style provides an excellent contrast to the snow-white tepals. The throat in white forms can be pure white, purple striped, or even deep purple. Once I almost discredited myself at a show of the Scottish Rock Garden Club when I thought that a pot with a deep purple-throated white crocus was incorrectly labeled as "*Crocus cartwrightianus*." I wanted to relabel it as "*C. mathewii*." At the last minute I noticed that the style was divided deep in the throat and this saved me from the dishonor. Otherwise, the plant was identical to *C. mathewii*. Later I gave to this form the cultivar name **'Purple Heart'** (Plate 95). Another crocus named 'Purple Heart' appeared in the trade some years later. It was selected by Antoine Hoog and has light lilac tepals with a deep purple throat. It strongly resembles *C. cartwrightianus* but the flowers close at night. According to Hoog, it is possibly a self-pollinated hybrid between *C. cartwrightianus* and *C. hadriaticus*. The division of the style helps to separate the white forms of *C. cartwrightianus* from *C. hadriaticus*, too. In the latter the stigmatic branches are shorter, more erect, and divide well above the throat.

According to E. A. Bowles (1952), *Crocus cartwrightianus* "flowers so freely in English gardens that it is very attractive on sunny slopes of a rock garden in October and November. Some of the white forms, especially if starred with purple lines in the throat, are particularly beautiful." In my climate, it is suitable only for pots where it increases very well, readily sets seed (especially if helped with extra hand-pollination), and forms large corms, even exceeding 3 cm in diameter in well-fertilized soil. It rewards my care with a long blooming and marvelous honey-violet fragrance.

Crocus hadriaticus Herbert

PLATES 96–98

Crocus hadriaticus (2*n* = 16) is another "pure Greek" species distributed in western and southern Greece. Its name is derived from the Adriatic region, although it is more strictly an Ionian species. To the east it is replaced by *C. cartwrightianus*. In the wild *C. hadriaticus* is found on stony or rocky hillsides of limestone or shale where it grows in open scrub or short grassland from 250 to 1500 meters.

The flowers are almost invariably white, rarely with a very faint lilac tinge. I saw such a specimen near Mistra at an elevation of 1100 meters. It was the only individual with markedly light lilac flower segments among a population with pure white flowers. The color of the flower tube can vary greatly from pure white to yellowish, brownish, or purplish staining extending a little on the tepal outside base. Several named cultivars are selected on the basis of flower segment shape and flower tube color.

In the type, ***Crocus hadriaticus* subsp. *hadriaticus***, the flower throat is almost invariably yellow and the filaments are 3 to 7 mm long and yellow. At a show in the United Kingdom I photographed a pot of blooms with very unusually colored throats: they were blackish changing to marred purple at the base of the outer sepals and similarly colored on the inner petals, only marred yellow. All in all, the plant looked like a typical *C. hadriaticus*. In **'Tom Blanchard'** the throat is quite small but brown with a wide diffused yellow edge.

White-throated plants growing near Mount Parnassus were separated as ***Crocus hadriaticus* subsp. *parnassicus*** (B. Mathew) B. Mathew (Plate 97). Their filaments are longer (8 to 11 mm) and white, but in all other aspects there is no great difference between these two subspecies, and recently such a separation has been questioned. Another form with slightly lilac-shaded flower segments and a white throat, is distributed at Parnon Range and was named ***C. hadriaticus* subsp. *parnonicus*** B. Mathew (Plate 98). Yellow-throated lilac forms are known as ***C. hadriaticus* subsp. *hadriaticus* f. *lilacinus*** B. Mathew.

My plants from the slopes of Mount Parnassus are slightly variable. They mostly have white, even grayish white throats, but in some samples the throat color can be described as creamy. In any case it is not quite yellow but also not pure white. Kees Jan van Zwienen (Netherlands) observed that 90 percent of the plants growing at Parnon Range above Agia Vasileois had white rather than pale lilac flowers; some may have been very pale lilac. None of the crocuses in that population had yellow throats ruling out subsp. *hadriaticus*. Subsp. *parnassicus* from further north (Mount Parnassus) also has white flowers with a completely white throat, and subsp. *parnonicus* is reported to occur on the summit of Monemvasía island but it was probably too early when I visited there as I found nothing but *C. goulimyi* and *Sternbergia lutea*. Because the line separating these subspecies is very thin, it is probably better to regard them as only varieties (B. Mathew, pers. comm.).

I have found *Crocus hadriaticus* to be easy to grow under cover, and it increases very well, producing large corms. Seedlings vary much in the color of the flower tube and tepal base. In milder climates it is a good grower in the open garden as well.

Crocus thomasii Tenore

PLATE 99

Crocus thomasii ($2n = 16$) from southern Italy and the western coast is the closest relative of *C. hadriaticus* but its flower color is usually a darker shade of lilac and the tepals are wider and rounder than those of its eastern neighbor. It is named after L. Thomas who collected it in southern Italy. On the opposite shore of the Adriatic Sea it grows in coastal mountains from the southern Velebit Mountains in the north as far as the Dubrovnik region in the south never penetrating inland more than 40 kilometers from the sea and growing in open dry rocky sites formed on limestone from sea level up to 1000 meters.

Flower color varies from light lilac to deep lilac with an invariably lighter or darker yellow throat which sometimes is orange or greenish yellow in the fairest forms but can always be described as yellow. The occasional lightest lilac-colored forms of *Crocus thomasii* can be distinguished from the darkest forms of *C. hadriaticus* by their leaves, which in *C. hadriaticus* are gray-green and very noticeably ciliated on the margins and on the keel while in *C. thomasii* they are green and glabrous or only slightly papillose on the margins. E. A. Bowles listed a rich lilac form with a reddish throat and a white margin on the outer segments. My darkest specimens have a somewhat dull yellow throat and the flower segments have darker veining.

I have not had much success with *Crocus thomasii*. All the stocks that I have received from various growers were infected with some fungal disease and perished within a couple of years. By careful selection and planting each corm into its own pot, I have in recent years been allowed to select healthier plants of this otherwise very beautiful species. Now my stock is built up, as the plants multiply quickly both by splitting and by seed. The plants need dry and hot summers and a well-draining soil mix. I have never tried it in the garden.

Crocus pallasii L. F. Goldbach

Crocus pallasii is the most widespread crocus of the saffron group and, given that, it is not surprising it is very variable. It was named after the German scientist P. S. Pallas who contributed to our knowledge of the flora in Crimea, from where *C. pallasii* was described. Brian Mathew separated the complex into four subspecies, combining under that name several species regarded by earlier botanists as distinct taxa. In fact this species is so variable that it is not easy to draw definite lines between subspecies, and recently Mathew (pers. comm.) stated that it probably would be better to regard *C. asumaniae* and *C. mathewii* as belonging to *C. pallasii*. I'm not so revolutionary, although the Crimean forms of *C. pallasii* subsp. *pallasii* are very different from the Turkish specimens of the same subspecies, so probably further subdivision is necessary. The division on the subgeneric level that follows is based on the length of the neck of the corm tunics, the width of the flower segments, and the color of the stigma.

Crocus pallasii subsp. *pallasii*

PLATES 100–104

The most widespread is *Crocus pallasii* subsp. *pallasii* ($2n = 14$) which occurs from Crimea on the northern coast of the Black Sea through Ukraine, Bessarabia (now Moldavia), Romania, Bulgaria, and the Aegean Islands and Turkey reaching Syria, Lebanon, and Israel. It grows on limestone or basaltic formations on open stony or rocky hillsides, in sparse scrub or steppe vegetation from 70 to 2820 meters.

Subspecies *pallasii* has the shortest neck at the top of the corm tunic compared with the other subspecies; it only slightly extends the apex and never exceeds 2 cm in length. Subspecies *pallasii* also has the broadest tepals, thus it is the most ornamental subspecies in the entire complex. It can sometimes be mistaken for *Crocus cartwrightianus*, but the two can be distinguished by the style which in *C. pallasii* divides above the base of anthers, but in *C. cartwrightianus* divides in the throat, thus resulting in much longer style branches.

I was rather perplexed when I received a marvelous crocus under the name *Crocus pallasii* **'Homeri'** (Plate 104) originally collected by Jimmy Persson on Chios Island (Greece), not far from the western coast of Turkey. Apart from its exquisite beauty enhanced by the almost blackish violet throat and black anthers (an uncommon feature in *C. pallasii*), it had such long style branches that at the first glance I even wanted to change its species name to *C. cartwrightianus*. Only after noticing that its blooms closed at night did I check the flowers more carefully and find that they had the very short filaments characteristic of *C. pallasii* and that the style was really divided above their bases.

The flower color of subsp. *pallasii* is very variable—from pure white to very deep purple with varying darker or lighter veining. Some forms are very beautiful and my favorites are those whose color is medium violet with a large white base which extends on the flower segments as white starry veins. One of the best forms was collected some 17 kilometers west of Konya in Turkey (BATM-460). I found a similar but more lilac-colored form near the Moca mountain pass in Cappadocia (RUDA-035). In other forms the throat is lighter or darker bluish with a darker star-shaped striping culminating in the aforementioned 'Homeri'. I have a pure white sample collected by a Czech collector from a roadside to the Altin yaila. The best flowers that I have seen are among the Crimean plants whose greatest advantage is the generally much wider tepals.

In Crimea *Crocus pallasii* subsp. *pallasii* grows on yailas with heavy clay soils. I have never seen it growing with *C. speciosus* and *C. tauricus* which prefer lighter soils. On the Kaya-Bash Heights the diversity of forms encompasses the entire spectrum of variability found in Turkey, but in general most plants have distinctly broader tepals although there are specimens with comparatively narrow tepals as well. In the best samples the tepal width exceeds 20 mm although the maximum width mentioned by

Brian Mathew is only 16 mm. Pure white forms exist with throat color varying from white to lighter or darker lilac with a grayish shade. With the exception of the pure white forms, the lilac-colored flowers usually have lighter or darker purple venation.

Crocus pallasii isn't difficult in cultivation, and in its best forms is a very valuable pot plant. It requires hot and dry summer conditions. In my collection it can be grown only under cover. In milder climates it can be planted in the open garden especially the forms from the western and northern parts of the area.

Crocus pallasii* subsp. *haussknechtii (Boissier & Reuter ex Maw) B. Mathew

PLATE 105

Crocus pallasii subsp. *haussknechtii* ($2n = 16$) has the longest fibrous neck at the top of the corm tunic, reaching up to 10 cm. The subspecies was named after the German botanist Heinrich Karl Haussknecht. It is distributed in western Iran, northeastern Iraq, and southern Jordan where it grows in dry fields and on rocky hillsides or in sparse oak scrub at 1300 to 2100 meters. The very long neck and somewhat more coarsely netted tunic allow easy identification even without flowers.

Plants from Jordan are fairly uniform and are white to pale lilac with distinct red–violet veins. The shape of the tepals varies from broadly ovate to narrow obovate. Most remarkable are the very erect and gray-green leaves.

I came across *Crocus pallasii* subsp. *haussknechtii* in Iranian Kurdistan on hard clay on open rocky slopes, not exactly on rocks but on moderately sloping sites where more soil had accumulated. In my samples the flowers are rather light lilac, actually nearly whitish, with slightly darker lilac venation though the color can vary. Generally the basic color remains whitish but the veins can be very dark lilac. The style branches are very deep red, widely expanded at the apex, and usually much shorter than the anthers. In general the tepal tips are more rounded than the tips in other subspecies, but this feature is not very consistent.

Since subsp. *haussknechtii* grows in such arid circumstances, in cultivation it requires very hot and dry summer conditions—at least from May to October—to flower well. Plants in my collection grow in pots and do bloom but not very abundantly. E. A. Bowles wrote that the subspecies sets seed readily.

Crocus pallasii* subsp. *dispathaceus (Bowles) B. Mathew

PLATES 106–108

Crocus pallasii subsp. *dispathaceus* ($2n = 14$) is more bizarre than beautiful. In the wild, the very narrow flower segments are never more than 7 mm wide. The name was given by E. A. Bowles who noticed the presence of a second basal spathe (prophyll)—a feature unique in the genus *Crocus*. Later when Brian Mathew dissected his specimens he didn't find this second spathe.

The flowers are reported as deep reddish purple with pale yellow style branches. Helmut Kerndorff collected it in 1987 near Aleppo in Syria. The collected samples mostly have deep violet narrow segments but some have deep vinous-purple and wider flower segments. Other populations included plants with dark violet flower segments, sometimes even brownish or with a gray-rose tinge. The style branches were always much shorter than the anthers and lemon yellow. No red styles were observed.

One of my samples was collected by Jimmy and Karin Persson in Turkey near Afyon, Kocatepe, on brown soil of igneous substrate at an elevation of 1870 meters. The style branches in this sample are orange, even red, and the tepals are light violet with darker veining becoming paler and nearly white in the throat. The flower segments are very pointed and narrow, making me think that the name is correct. The style is very short, its branches end at the middle of the curving yellow anthers. In this last feature it well matches the description given by E. A. Bowles (and the pictures of Robert and Helmut Kerndorff in the *Alpine Garden Society Bulletin* 64 [3]) but not in the color, which in Bowles's words, was "the deepest vinous purple of any Crocus." In the identification key I used the yellow stigmatic branches to separate this subspecies from the other subspecies, but if the stigmatic branches are orange, then curved anthers enclosing them can help to distinguish this subspecies from subsp. *turcicus*.

Crocus pallasii subsp. *dispathaceus* is not well known in the wild nor in cultivation. Even its natural distribution area is not precisely determined; it was found in northern Syria and (possibly) in southern Turkey, at 300 to 2000 meters, on limestone.

I grow this subspecies in the same way as subsp. *haussknechtii* giving a dry and hot summer rest from at least the end of April through the end of October. According to Helmut Kerndorff, it needs sandy soil with good drainage like all the species from semidesert areas; otherwise it seems not to be a difficult plant in cultivation. Samples from Ma'arrat and Nu'man seem to be more vigorous than those from Aleppo.

Crocus pallasii* subsp. *turcicus B. Mathew

PLATES 109 & 110

Crocus pallasii subsp. *turcicus* ($2n = 12$) is the fourth subspecies determined by Brian Mathew. It occurs in southeastern Turkey, Syria, and Lebanon, where it grows in very dry regions on rocky places among steppe vegetation at 600 to 1700 meters.

My samples from Lebanon and from Syria have been grown from seeds collected by Jim and Jenny Archibald. The Lebanese plants are more variable in color—from very pale lilac with slightly darker veins to densely lilac-veined on a very pale background that on the whole gives a richly colored appearance. The Syrian plants are very similar to the pale form from Lebanon, although their stigmatic branches are very short and end at the middle of the curving anthers, which would have forced me to think about subsp. *dispathaceus* if the tepals had been narrower.

The distribution area clearly indicates that this subspecies needs hot and dry summer conditions. The greatest problems during the winter are the lack of sunshine and the dull and wet weather.

Having now reached the end of this review of *Crocus pallasii* and its subspecies, I have to admit that I'm not very satisfied with this part of my work. Our knowledge about this species *sensu lato* still is too limited. Quite often, wild-collected corms or seeds are automatically labeled according to where they are found, so that, for example, material collected in Iranian Kurdistan, is labeled as subsp. *haussknechtii* simply because that subspecies is known to grow there. At the moment I would prefer to divide the whole *C. pallasii* complex into two groups—one with subsp. *pallasii* consisting of plants without or with a very short corm tunic neck and the other consisting of the three remaining subspecies whose tunic necks are up to 10 cm long. From a gardener's perspective, forms of subsp. *pallasii*, especially those from Crimea and Bulgaria with wider flower segments, are more spectacular and easier to grow than forms of the other subspecies.

Crocus naqabensis Al-Eisawi & Kiswani

Recently described from Jordan, *Crocus naqabensis* ($2n = 14$) seems to be similar to *C. pallasii* subsp. *pallasii*. It is reported as having a reduced tunic that does not form a neck and a white glabrous throat which in *C. pallasii* is pubescent. In the original description *C. naqabensis* is compared with *C. moabiticus* and *C. cartwrightianus*, both of which have long tunic necks, a pubescent throat, and longer style branches. Furthermore, *C. moabiticus* starts to form leaves only when flowering has ended. Very little information is available about this species and I have no knowledge of it being cultivated.

Crocus asumaniae B. Mathew & T. Baytop

PLATES 111–113

Crocus asumaniae ($2n = 26$) was first collected by Asuman Baytop from the University of Istanbul and named after her in 1979. It is closely related to *C. pallasii* but is less leafy. If *C. pallasii* usually has 10 to 17 leaves, *C. asumaniae* only has 4 to 7 leaves although on well-grown multiflowering corms the number of leaves can reach 13. The corm tunics of *C. asumaniae* are not as distinctly reticulated as those of *C. pallasii*.

Crocus aumaniae was found originally not far from Akseki near an old road to Beyşehir, where it grows in openings in oak (*Quercus cerris* and *Q. coccifera*) scrub on stony clay ground with limestone outcrops at 900 to 1250 meters. I found it on the *locus classicus* near an abandoned old road where it grew densely between heaps of rubbish on every more or less open spot among rather high shrubs. Later I found it quite far to the west shortly below Karaova Beli pass, where it bloomed on a roadside slope at forest edge. This confirmed much wider distribution of *C. asumaniae* than supposed earlier.

White is commonly given as the color of *Crocus asumaniae*. Surprisingly, all my samples of this species, which I acquired from various sources, have pale lilac flowers varying only in the intensity of purplish striping in an otherwise grayish white throat. There can be some dark stripes on the outer base of the flower segments which sometimes stretch down the flower tube. The style divides well above the base of the anthers into three reddish orange branches which well exceed the tips of anthers.

In 2007 I obtained a pure white form from Joachim Sixtus and immediately fell in love with it for its long bright red stigmatic branches which contrast excellently with the brightest white of the pointed perfectly formed flower segments. The throat of this white form is light yellowish, and the flower tube is a somewhat grayish shade (Plate 113).

Crocus asumaniae is a good grower and increaser and sets seed perfectly. Now, it is well established in cultivation and is readily obtainable. In summer it needs a dry and warm rest. I have never tried it in the open garden.

Crocus mathewii Kerndorff & Pasche

PLATES 114 & 115

Crocus mathewii ($2n = 70$) is closely allied to *C. asumaniae* but in its type form is easily distinguished by its prominent bright purple throat that puts it among the showiest and most desirable plants of every crocus collection. It was discovered in the autumn of 1992 by German plant explorers Helmut Kerndorff and Erich Pasche, who found it in the Lycian Taurus Mountains of the province of Antalya in southern Turkey growing at 400 to 1100 meters. It was named in honor of Brian Mathew.

Initially known only from a few records, *Crocus mathewii* was found later to be more widely distributed and quite variable in the size of the inner purple throat and the color of the flower tube. In the best forms the flower segments are wider, and the throat is large and enhanced by the deep purple flower tube. In other forms the tube is white with only a small grayish blue "tongue" on the base of the exterior side of the sepals, and the flower segments are more pointed. Flowers are unscented. In the wild it grows on gentle slopes sparsely covered by dwarf spiny shrubs as well as in grass in more open places.

Alan Edwards selected one very good form which he named **'Brian Mathew'** (Plate 115). The flower has a very large deep purple throat at least one third of the tepal length and a purple flower tube extending upwards and making a comparatively large deep purple blotch at the outside base. The flower segments are wider and rounder than in most of the unnamed forms. At the Gothenburg Botanical Garden an excellent form was selected and named **'Karin'** by Karin Person who collected the original plant near Ovacik on Baba-Dag ridge. Its flowers are very nice light lilac with a large deep purple throat.

Albino forms without a purple throat are sometimes found among the typical plants, and then can arise the problem of how to separate these from white forms of the closely related *Crocus asumaniae*. Sometimes the albinos have a slightly violet color on the flower segments, making them even more similar to *C. asumaniae*. Peter and Penny Watt found populations where the flower color varied widely from white to lilac and where only a few plants had lilac throats; in some stands purple-throated plants are scarce, in others predominant. Some botanists now think that *C. mathewii* may need to be united with *C. asumaniae*.

Crocus mathewii can hybridize with *C. pallasii* where they meet in the wild (at

Fethie). There are forms with similarly purple-colored throats in other species as well and at times the identification is not easy.

Crocus mathewii is easy to grow and readily sets seed. Seedlings start blooming in the third year. Plants increase well vegetatively. The largest clump seen by me in nature was formed from 12 vegetatively formed corms. I grew this species for a few years in the open garden during my pre-greenhouse era, but have since relocated my plants under cover for safety reasons.

Spring-Blooming Crocuses

Spring in Latvia is usually associated with crocuses, not snowdrops which sometimes go unnoticed with their very small white silent bells. It is the crocuses—with their bright colors, the bees buzzing over their blooms in search of the season's first nectar—that announce spring's arrival and ask, Where are you, gardener, with a spade and rake, seed bags, catalogs, and notebooks under your arm? Come and start the gardening year!

It is among spring crocuses that we can find the greatest variability of color, flower shape, and blooming time. In general spring crocuses are easier to grow and, with a few exceptions, most of them can be grown outside; only a few require specific growing conditions. Spring-blooming crocuses let winter, the most difficult season, pass while they lie sleeping in the ground without leaves. In the case of species that need hot and dry summers, the corms are lifted at the end of vegetation and the required conditions easily provided in the bulb shed.

While writing this chapter, I though long and hard about how to best group the spring-bloomers to make their identification easier. I borrowed several groupings from E. A. Bowles's *Handbook*, though many new species have been described since that publication was revised. Some of my groups are natural, for example, the annulate crocuses. Others are very artificial, as for example, when I group species by flower color—yellow or blue/white—excluding the ones discussed in other chapters. The basic principle guiding me was to make all the spring-blooming crocuses easier to identify. Therefore, the keys sometimes include species reviewed in other chapters. I tried to avoid basing the keys on knowing the exact origin of the plant though in some cases, particularly in series *Biflori*, it was not easy.

The Fellowship of the Rings—The Annulate Crocuses

This group includes the species which belong to Brian Mathew's series *Biflori*, namely, those whose tunics split in the lower part of the corm into a series of narrow overlapping rings. Although straight-edged below, higher up they are equipped with stiff teeth of various lengths and this feature is sometimes used to separate taxa, but was proved by several botanists as too inconsistent. In some species the basal rings are indistinct or even absent, but the overall shape of the corm tunics remains membranous smooth and only in a few species the splitting is vertical at the base but without any fibrous structures. Although these are not strictly annulate, I included them in this chapter as they all belong to the same series *Biflori*. The tunics can be fine, membranous or hard and coriaceous.

This group includes most of the newly described taxa, resulting primarily from the work of two dedicated crocus explorers from Germany—Helmut Kerndorff and Erich Pasche—who each spring spend much time in the mountains of Turkey searching for new forms. Their endeavors have been rewarded with many new discoveries. Variability within the species and subspecies sometimes is so great that it is almost impossible to make a satisfactory key. On the other hand these crocuses are very popular with gardeners because of their early blooming, bright colors, and the relative ease of cultivation. A lot of cultivars have been raised through clone selection and hybridizing and in numbers of cultivars it is the richest group among crocuses.

In his monograph Brian Mathew joined many of the annulate crocuses previously regarded as separate species under a single species (*Crocus biflorus*) with many subspecies. Some isolated or very distinct-looking species were not included, although recently Brian wrote to me that he tends towards the opinion that probably it would be better to regard all the species with annulate tunics as subspecies of *C. biflorus*, even *C. chrysanthus* and *C. cyprius*. All botanists can be divided into two groups—the lumpers and the splitters. The former tend to join several different taxa in a larger one, the latter are inclined to split species in many smaller kinds. As for me I would call myself a "moderate splitter." I accept the subspecies status but prefer to deal with species.

Most of the annulate crocuses are spring blooming; those which bloom in autumn were treated in the previous chapters. The key that follows, like the main keys in other chapters of this book, includes most of the subspecies, but *Crocus biflorus* is an exception. It now has so many recognized subspecies that to include them would make the key too long and more difficult to use. Instead, an identification key for the subspecies of *C. biflorus* is provided at the beginning of the corresponding review.

1. Flowers yellow
    2. Leaves 3–5 mm wide .... *C. almehensis*
    2. Leaves less than 2.5 mm wide
        3. Flowers up to 1.5 cm long, filaments 2–3 mm long
            4. Flowers usually pale yellow .... *C. danfordiae* subsp. *danfordiae*
            4. Flowers bright yellow .... *C. danfordiae* subsp. *kurdistanicus*
        3. Flowers 1.5–3.5 cm long, filaments 3–6 mm long, usually bright yellow .... *C. chrysanthus*
1. Flowers white or blue
    5. Filaments orange-red to red
        6. Anthers black .... *C. hartmannianus*
        6. Anthers yellow .... *C. cyprius*
    5. Filaments yellow with a black stain at base .... *C. pestalozzae*
    5. Filaments yellow without black at base
        7. Corm tunics with more or less distinct basal rings
            8. Flower segments up to 1.5 cm long, filaments 2–3 mm long .... *C. danfordiae*
            8. Flower segments 1.5–3.5 cm long, filaments 3–6 mm long

9. Basal rings with long teeth equal or even longer than the width of proper ring .... *C. tauricus*
9. Teeth on edge of basal rings shorter
    10. Throat yellow .... *C. biflorus*
    10. Throat not yellow
        11. Flowers white inside .... *C. biflorus*
        11. Flowers blue inside .... *C. adanensis*

7. Corm tunics membranous, splitting at base without distinct rings
    12. Anthers before dehiscence grayish or brownish green, at least connective dark, leaves 6–13 .... *C. leichtlinii*
    12. Anthers yellow
        13. Tunics papery
            14. Leaves gray-green, throat yellow or white
                15. Flowers blue with distinct veining, throat yellow .... *C. aerius*
                15. Flowers blue without distinct veining, throat white .... *C. adanensis*
            14. Leaves grass green, flowers blue without distinct veining, throat yellow .... *C. paschei*
        13. Tunics tough, flowers light blue with a dark median zone (stripe) on backs of outer segments .... *C. kerndorffiorum*

Crocus chrysanthus (Herbert) Herbert

PLATES 116–126

Crocus chrysanthus is one of the most popular crocus species. Its bright yellow flowers shine in the garden from afar. Under this name are grown cultivars in a wide range of colors. Many of them are hybrids between several variants of *C. chrysanthus* and *C. biflorus* and I don't think that it would be correct to apply the epithet "*Crocus chrysanthus* cultivars," particularly if they are not yellow in color. Among the synonyms of *C. biflorus* and *C. chrysanthus* exist the name *C. annulatus* and I think that for the cultivars of this group it would be best to apply the name *C.* ×*annulatus*, in such a way characterizing the shape of corm tunics and accenting their hybrid origin.

The species name means "golden flower" and really the flowers of *Crocus chrysanthus* are almost invariably bright golden yellow. Given the wide range of chromosome numbers in this species—from $2n = 8$ to $2n = 20$—it is no wonder that variability within the species is very great and such features as the pubescence of the throat, the position of style branches, the number of leaves, and so forth, vary widely throughout the range. It wouldn't be surprising if several species that are indistinguishable without studies on a sub-molecular level were to be joined under this name.

In the wild *Crocus chrysanthus* grows in an area that stretches from central and southern Turkey through central and northern Greece to former Yugoslavia, Albania up into central Bulgaria in the north where it usually can be found in open places in short grassland or sparse scrub from sea level up to 2200 meters. Flowers generally are bright yellow but in southwestern Turkey forms with brown suffusion or stripes on the outer flower segments are quite common. Albinos are rare but in some populations in the European part of Turkey they can be quite common. Plants with a beautiful purple flower tube are common in southwestern Turkey (Plate 118). The style is trifid and its color usually can vary from yellow to bright orange-red. The anthers normally are yellow, or yellow with black-colored basal lobes. In very rare occasions the anthers can have a black connective or be entirely black.

In some places *Crocus chrysanthus* and *C. biflorus* grow side by side though usually they are separated ecologically—*C. chrysanthus* prefers drier places at lower elevations while *C. biflorus* grows in damp meadows and flowers near melting snow. In some places they can be found in mixed populations and there cross-pollination occurs, especially if both are represented by forms with an identical chromosome number.

It is not easy to distinguish the pure white form of *Crocus chrysanthus* from the similarly white *C. biflorus* subsp. *weldenii*. I have two white "*chrysanthus*"—one is of an unknown origin whose only somewhat warmer shade of white allows me to include it among *C. chrysanthus* varieties. The other white form was originally collected in Macedonia near Kičevo and its anthers have pleasingly gray tips on the basal lobes, thus clearly indicating the affiliation with the large *C. chrysanthus* family (Plate 119).

I photographed a very unusual form in the Ulu-Dag mountains in northwestern Turkey where it grew in a small meadow in pine forest just near the gates of the National Park among *Crocus biflorus* subsp. *pulchricolor* and *C. herbertii*. It was snowing when I found this beauty, so the flower was closed and I saw only its outside. The flower tube was golden yellow changing into wide bright yellow stripes halfway up the otherwise creamy white tepals. I haven't seen similar coloring in any other crocus.

A very popular form of *Crocus chrysanthus* was raised in the Van Tubergen nursery when it was operated by the Hoog family. It was first offered in the trade as **'Uschak Orange'** in 1973 at a price that was 20 times higher than the average price of *C. chrysanthus* cultivars. My first plants had orange stigmas and were the first to bloom in my collection. Later, when the Hoogs sold the Van Tubergen nursery, I got a few additional corms from Michael Hoog, but they, although of the same color and flowering time, had yellow stigmas. Both are fertile and set seed abundantly, resulting in plants with stigmas of both colors. I would have preferred to keep as 'Uschak Orange' the plants with orange stigmas. This form has the chromosome number $2n = 20$ and was reportedly collected near Uşak, Turkey. Specimens with the same chromosome number are also found in Yugoslavia and Greece.

Particularly beautiful are forms with a very contrasting black color in the golden yellow blooms. When the variety **'Sunspot'** (Plate 120), raised at Pottertons Nursery, appeared for the first time it shocked everyone by its shining black tips of the short stigmatic branches. It turned out to be a very floriferous and excellent grower. It can produce as many as 11 flowers from a single corm.

In *Bulbous Plants of Turkey and Iran* (Sheasby 2007) is a picture of a *Crocus chrysanthus* specimen with pure black anthers from the Honaz-Dag in western Turkey. This picture motivated me to search for similar plants on the Gembos yaila near Akseki where *C. chrysanthus* grows alongside *C. biflorus* subsp. *isauricus*. The latter often has black connectives and on the Gembos yaila both species hybridize. My goal was to find similar plants and to cross them to obtain a flower with both a black stigma and black anthers. During the day-long search I found two specimens with black stigmas (Plate 121), not as shiny as in 'Sunspot', but red right at the branching point and gradually changing to black at the tips. The best of the samples had broken up through asphalt near a roadside drinking fountain, and it was a wonder that it was not trampled down. It took me almost an hour to carefully set it free. Fortunately the corm was close to the surface. I didn't see any crocuses with black anthers at the site, but I did find two specimens with a distinctly black connective on the anthers (Plate 122). Unfortunately in both variants the flower exterior was not as pure shining golden yellow as in 'Sunspot' but had a grayish flush on the sepal backs in one specimen and brownish stripes in the other. I presume that these can be crossed because they are from one population and must have identical chromosome numbers.

An excellent brownish form was selected in the Gothenburg Botanical Garden from plants collected by the garden's team near Gündoğmuş in the same Akseki area. Later this form was named **'Gundogmus Bronze'** (Plate 123). It is exceptionally deep golden yellow inside and, on the outside, the back of the outer segments is entirely covered with deep brown speckling. The plant is somewhat similar to the form described as *C.* ×*bornmuelleri* from near Skopje with $2n = 13$, indicating that most possibly it is of hybrid origin.

Hybrids between *Crocus chrysanthus* and *C. biflorus* are among the most popular spring flowers. Because their assortment changes constantly and the most popular varieties can be acquired at low prices through bulb catalogs and in supermarkets, I won't list them here. An excellent article by Jacobsen et al. (1997) includes a key, short descriptions, and nomenclatural notes on 49 identified and 6 unidentified crocus cultivars of this group.

I have sown open-pollinated seeds of these famous garden cultivars. The resulting seedlings varied greatly. Some of them I even named as they were very good compared with the parent cultivars. From these I want to mention only very few. One of the best has very large, perfectly rounded purest white flowers, which open widely in the sun and which I named **'Snow Crystal'** (Plate 124). It has grayish purple speckles on the flower tube and similarly colored long and wide "tongues" on the outside base of the outer segments. The throat is grayish or greenish yellow. The relationship between this beautiful variety and *Crocus chrysanthus* is confirmed by the gray tips of the basal lobes of the anthers.

Another of my open-pollinated hybrids, **'Goldmine'** (Plate 125), is the first double-flowered crocus. Currently it is the only more or less constantly double (actually semi-double) cultivar, producing approximately 80 percent double flowers from large corms. The flowers have double or even triple the number of tepals. The cultivar's only fault is that the corms must be very well-grown and of good size, otherwise the flowers will have the normal number of tepals. I named this hybrid not only for its brilliant golden yellow color but also for the financial return I get from it.

A number of other crocus species occasionally produce semidouble flowers. I have a semi-double autumn-flowering selection of *Crocus melantherus* raised by John Fielding, but the stock is still too small to judge the consistency of this feature. *Crocus tommasinianus* 'Eric Smith' is reportedly semi-double but it is less constant and in my experience it gives eight-tepalled flowers only if left undisturbed for a second year and well fed during the previous season.

An excellent hybrid was raised by Latvian tulip breeder Juris Egle. The outside of the flower segments is nicely violet striped and feathered over the white background color. When flowers open they have a very rounded appearance with the inside segments blue at the tips, becoming lighter to white towards the base, and with a very deep yellow throat. It was named **'Jūrpils'** (Plate 126) after a small village in Latvia.

In the garden *Crocus chrysanthus* is very easy, set seeds well, and even is self-sowing but I never got seeds from white-flowering stocks of *C. chrysanthus*. Seed pods often split when they are at least partly in the soil. I take pot-grown plants outside during the summer, where it is cooler than in the greenhouse.

Crocus almehensis C. D. Brickell & B. Mathew

PLATES 127 & 128

Crocus almehensis ($2n = 20$) superficially is very similar to *C. chrysanthus*. It is named after the village of Almeh in Gulistan National Park in northeastern Iran. Although in flower *C. almehensis* resembles *C. chrysanthus*, both species are easy to distinguish by their leaves, which in *C. chrysanthus* are up to 1.5 mm wide and flat or only very shallowly channeled whereas in *C. almehensis* they are almost V-shaped and up to 5 mm wide. The flower tube is purplish striped. The outer base of the flower segments in my plants is marked with short deep mahogany-colored stripes but they don't run up the sepal back.

John Ingham, who visited Almeh in spring 2007 during the peak of *Crocus almehensis* blooming season, took pictures of forms with a narrow long brown stripe along the midvein of the outer tepals; in some samples the stripe reaches the tip of the flower segments (Plate 127). The flowers can be pure yellow without any stripes even at the base, as is the case for a specimen I recently obtained from the Gothenburg Botanical Garden (Plate 128). Ann Borrill who visited the *C. almehensis* locality in 2002 reported specimens with very dark, almost black stripes dominating among crocuses blooming by the melting snow; few were without any markings.

This crocus comes from high elevations where it flowers near snow melt on upland pastures and in steppe vegetation. Spring of 2008 came very early, so that during my visit to Gulistan National Park the flowering was long over.

Although not very difficult in cultivation *Crocus almehensis* increases slower than *C. chrysanthus* and is still extremely rare in cultivation and known only in a few collections. I keep it dry and warm during the rest period. It seems that *C. almehensis* is self-sterile, so it is necessary to cross-pollinate it with other specimens to get seeds.

Crocus danfordiae Maw

PLATES 129 & 130

The third species in series *Biflori* that can have yellow flower segments is *Crocus danfordiae* (2*n* = 8). This species is unique among crocuses as it is the only one in which a population can contain equally mixed specimens with a yellow, blue, and white flower color. The species is named after Mrs. C. G. Danford who first collected it in southern Turkey. It grows on open hillsides in short grassland in sparse scrub from 1000 to 2000 meters. Although in some places it grows close to *C. chrysanthus*, both species are easily separable by the pale color and the small size of *C. danfordiae* flowers. Even in the darkest yellow forms it lacks the brilliancy of its larger cousin. I came across some very deep yellow specimens of *C. danfordiae* between Beyşehir and Huğlu and just the small size allowed me to distinguish it from forms of *C. chrysanthus* growing in close proximity.

The flowers are up to 1.5 cm long and apart from the color are very uniform in overall appearance. The sepal outsides usually are finely speckled with minute gray or purple dots, somewhat similar to those of *Crocus biflorus* subsp. *punctatus*. Like in *C. chrysanthus*, the basal lobes of the anthers are often grayish: sometimes the dark zone is very tiny, sometimes very prominent. The yellow to orange style branches usually reach only the middle of anthers.

Although I can grow this very vigorous growing plant in the open garden, it is too small and it does much better under cover where it flowers very freely. In pots the small size of individual flowers is well compensated for by the abundant flowering. *Crocus danfordiae* is a good increaser and readily sets seed. During summer it requires dry and warm conditions.

Crocus danfordiae* subsp. *kurdistanicus Maroofi & Assadi

In 2002 a new crocus from Iranian Kurdistan was described with bright yellow though very small flowers and, because they closely resembled flowers of *Crocus danfordiae* in size, the new form was named *C. danfordiae* subsp. *kurdistanicus*. It differs from *C. chrysanthus* in flower size and from *C. almehensis* in size and in the pure yellow color. The latter feature is of little importance for there are *C. almehensis* forms with pure yellow flowers, too. There is very little information about this new crocus and it isn't introduced in cultivation yet. Its great distance from all known relatives supports the assumption that it can be a separate taxon.

Crocus pestalozzae Boissier

PLATE 131

Crocus pestalozzae (2*n* = 28) is another species with comparatively small flowers and is named after the Italian doctor F. Pestalozza, who botanized in Turkey. It is quite unique among the annulate crocuses by the rigid bright green cataphylls rising from the ground. Another unusual feature is the blackish or brownish stain at the base of the otherwise bright yellow filaments. This is a very constant feature and enables an easy identification of this species. E. A. Bowles commented on this feature, "These minute black specks at a first glance look like tiny pellets of soil fallen into the throat of flower." Only quite recently was a new subspecies of *C. biflorus* discovered with a violet-shaded base of filaments (subsp. *ionopharynx*), but its flowers are basically deep violet-blue.

In the wild *Crocus pestalozzae* grows at low elevations and has quite a limited distribution in the European and Asian parts of northwestern Turkey where it can be found in short grassland and bare stony places in low hills (at 90 to 200 meters). It prefers acid soils.

Crocus pestalozzae is rarely collected by travelers in the wild as it blooms very early, when there are few other flowers. In some locations it grows with the similarly colored *C. biflorus* and the larger blooms of latter usually attract more attention of bulb collectors.

Flower color can be white or blue. The white form occurs in both European Turkey and in the adjacent northwestern Asiatic part of Turkey. The blue form is most common in cultivation, though in nature it is very rare and its provenance was not accurately known (referred to as collected "near Constantinople"). Very recently it was rediscovered in the wild by Ibrahim Sözen (Turkey). It was so rare that he saw only a few plants after walking a whole day. The location was near a fountain where cows and sheep drink in a meadow that was covered by water from melting snow. At present both forms are only rarely available in the trade, but only the white-blooming specimens have a known wild origin. According to Brian Mathew the cultivated blue-colored form is a very vigorous clone "almost certainly derived from the plants introduced by Hoog and Barr." Phylogenetic studies have confirmed that both color forms belong to the same species.

I can grow *Crocus pestalozzae* only under cover. In fact, before I had a greenhouse, I lost both color forms very soon after obtaining them. I like the white form better but have only been able to get the blue form, which does well in pots. I keep the pots dry during the summer. The blue form quickly multiplies by division and is referred to by Alan Edwards as a good seeder in most years.

Crocus cyprius Boissier & Kotschy

PLATES 132 & 133

Especially beautiful, regardless of the comparatively smaller size, is *Crocus cyprius* ($2n = 10$). It is a very narrow endemic known only from the western part of the Troodos Mountains in Cyprus, where it grows from 1000 to 2100 meters on open rocky slopes, in sparse scrub and open pine woodlands. There it flowers by melting snow from February to April.

The flowers are fragrant; the basic color ranges from white to lilac. Only lilac specimens are represented in my collection. All of them have a very dark, sometimes almost blackish purple basal "tongue" on the outer flower segments. In some specimens the "tongue" goes up two thirds of the sepal length. Particularly beautiful is the very bright orange throat which sometimes shines through the tepals even when the flowers are closed. The filaments are bright orange, in some forms almost red, and this helps to separate it from other species. The leaves usually develop after flowering, their green tips showing up shortly before the end of blooming. The corm tunics are rather thin and papery, splitting lengthwise, and the basal rings are not very distinct and somewhat divided vertically.

Crocus cyprius is still very rare in cultivation and is grown in only a few collections. It needs plenty of water at flowering time and a hot and dry summer, so it is suitable only for pots. Insufficient watering in early spring can seriously reduce the crop of corms though otherwise it increases quite well and readily sets seed. The corms are somewhat sensitive to drying out when stored in a bulb box without soil. I try to replant them immediately after lifting, paying particular attention that the lifted bulbs are not exposed to sun while being harvested. Once I lost most of my *C. cyprius* corms because of that.

Crocus hartmannianus Holmboe

Crocus hartmannianus ($2n = 20$) is similar to *C. cyprius* and shares with it the red color of filaments. It was named after E. Hartmann who originally collected it near the easternmost peak of the Troodos Mountains in Cyprus. It grows at 800 to 1000 meters on dry rocky slopes in *Cistus* scrub and in sparse pine woodland.

The flower color resembles that of *Crocus cyprius*, but the outside of the flower segments is more or less striped. The variation among individuals can be great, ranging from a broad purple-black stripe to almost completely dark outer sepals, even with feathering. The background color always is pale mauve with a bright orange throat. Both species are easily separated by the black anthers, which add beauty to *C. hartmannianus*, although its tepals are narrower and the overall appearance is not very imposing. Another feature which allows distinguishing between the two species is the development of leaves. In *C. hartmannianus* their development starts before the flowers and sometimes they equal or often surpass the flowers by the end of blooming. Their corm tunics are also different; those of *C. cyprius* have basal rings, but the papery tunics of *C. hartmannianus* split into many parallel fibers.

Crocus hartmannianus is very rare in the wild as well as in cultivation, and is known only in some collections. In general it is even more difficult than *C. cyprius* and suitable only for pots which must be kept warm and dry during the summer rest. Apart from the black anthers which are my weak point, I like *C. cyprius* better because of its more saturated and better-formed flowers, and it is one of my real favorites.

Crocus leichtlinii (Dewar) Bowles

PLATES 134–136

Crocus leichtlinii ($2n = 20$) is another representative of series *Biflori* without basal rings at the corm base. It was named by E. A. Bowles after German nurseryman M. Leichtlin who sent its corms to the Royal Botanic Gardens, Kew. In the wild it occurs in southeastern Turkey where it grows on open rocky places in soils of volcanic origin at 1100 to 1800 meters.

The open flowers seem almost white but in bud they have a nice very light bluish flush that is deeper near the sepal midribs and covers the entire sepal outside and becomes somewhat greenish at the base. The throat is large, yellow to bright yellow, even orange-shaded, and shines through the tepals. In the best forms the tepals are wide and round. The anthers are grayish or brownish green, sometimes only the connective is dark, but when the anthers dehisce the pollen grains hide the dark color. The style branches reach the middle of the anthers, rarely equal or even slightly exceed their tips. The flower tube is white turning yellow near the throat. According to Brian Mathew, the flowers are heather-scented.

Crocus leichtlinii is still a rarely cultivated plant and can be found in a few collections. In the wild it grows in harsh conditions with cold winters and very hot and dry summers, so it is suitable only for growing in pots under cover. My experience with this species is very limited, as I only recently obtained a specimen of it. E. A. Bowles writes that it sets seed in cultivation.

Crocus kerndorffiorum Pasche

PLATE 137

Crocus kerndorffiorum (2*n* = ?) was discovered quite recently and named by Erich Pasche in honor of the crocus enthusiasts Helmut and Robert Kerndorff who discovered it in the eastern Taurus Mountains in southern Turkey. It grows at 900 to 1600 meters on unconsolidated screes and is thinly scattered in oak, juniper, and pine forest, being very local and very rare. It grows particularly among north-facing large rocks and in rock crevices but can be found on north- and northwestern-facing slopes in unconsolidated screes or gullies.

The corm tunics of *Crocus kerndorffiorum*, like those of *C. leichtlinii*, do not split into horizontal rings at the base. They are membranous and split into narrow parallel strips, whereas the tunics of *C. leichtlinii* are stronger, leathery, and split in triangular teeth at the base. Because little is known about *C. kerndorffiorum*, it isn't easy to evaluate its variability. The forms known to me look very similar to *C. leichtlinii*, but lack its greenish shade. The flowers of my *C. kerndorffiorum* stock are light blue, the throat is lighter yellow and smaller in size, but the central stripe on the back of the sepals is longer and much more prominent than in *C. leichtlinii*. At the base is a wide dark zone which covers the upper part of the flower tube and fades downwards. Both species are easily separable because the anthers in *C. kerndorffiorum* are yellow with a gray-green margin (which in general can be called yellow) while in *C. leichtlinii* they are much darker and can be described as blackish in overall appearance. The style in *C. kerndorffiorum* usually exceeds the anthers and sometimes is quite unusually colored—at the base it is yellow, higher up, below the branching, it becomes dark brown and the branches are deep red but I haven't noticed this feature in my plants (HKEP-9010). In *C. leichtlinii* the style branches usually are shorter than the anthers or reach their tips and the color is yellow or orange throughout. Both are well-isolated geographically, although in the new phylogenetic tree they are placed side by side.

Although *Crocus kerndorffiorum* is still very rare in cultivation, the first impression is that it is not a very difficult plant to grow in pots and is more floriferous than in the wild. It needs rather cool but dry summer rest. Kerndorff and Pasche wrote that "although it needs a dryish summer rest, it dislikes being baked; instead it should be kept in a rather cool corner or in pots." They recommend a 5-cm thick mulch of pine needles to keep the conditions fairly stable and to protect the plant from extremely high temperatures in summer. My plants regularly produce two to four flowers per corm and set seed well without extra hand-pollination. Seeds germinate very well.

Crocus paschei Kerndorff

PLATE 138

Crocus paschei (2*n* = ?) originally was described by Helmut Kerndorff as belonging to series *Flavi* but subsequently Brian Mathew assigned it to series *Biflori*. Phylogenetic studies show that it is very close to *C. adanensis*, another species listed by Mathew in series *Biflori*. Therefore, I decided to include it in this chapter although both species form a sister group to the species of series *Flavi*.

Crocus paschei was named after Erich Pasche. In nature it occurs in open forests, scrubland, and maquis as well as in adjacent fields in the Taurus Mountains of Turkey from 700 to 1400 meters. In some places it can grow quite abundantly.

In the original description *Crocus paschei* is characterized as similar to *C. antalyensis* in flower color, though it is paler and lacks the conspicuous stripes on the sepal backs. The throat is described as yellow with a white band, but in my samples with a pale yellow throat, the white zone is very indistinct. In the darker-colored samples, the white zone is more prominent though much diffused and more pronounced on the petals. The sepal outside usually is silvery or light violet and more speckled; sometimes a diffused indistinct striping can be seen. The perianth tube is white, and toward the flower base has occasional lilac speckling. The filaments and anthers are yellow, the style branches bright orange and well exceeding the anthers, only rarely equal to them. In my plants the flower segments are somewhat pointed and narrow, although I have seen other plants with more ovate segments. The leaves are grass-green and frequently overtop the flowers. The corm tunics are papery, occasionally splitting lengthwise at the base.

The species's phylogenetic neighbor, *Crocus adanensis*, has ringlike basal tunics and a white flower throat. Sometimes it is not easy to separate the two species as some forms are quite similar. In *C. paschei* the throat can be light-colored but usually is yellow-shaded (although I have a few specimens with almost no yellow in the throat, but then it isn't white either, it is more grayish diffused on tepals, without a distinct border) and the leaves are grass-green. In *C. adanensis* the throat is white and well-defined from the rest of the segments (somewhat starry but clearly bordered from the lilac-colored part), the leaves are gray-green, and there is a greater contrast between outer and inner tepals. I have never had a problem distinguishing the samples I grow.

My first impression of *Crocus paschei* is that its flowers lack substance, are somewhat too pale and hidden among leaves at the peak of blooming. It proved to be somewhat less vigorous than its close relatives. Brian Mathew states that it is not difficult in cultivation and, although still rare, is well established in specialist collections. I grow it in pots, and it sets seed for me which germinate well.

Crocus aerius Herbert

PLATE 139

The last species with indistinct basal rings but in overall appearance clearly belonging to series *Biflori* is *Crocus aerius* ($2n = 22$). Its name means "aerial" and refers to the high mountain habitat in northeastern Turkey where it grows in alpine meadows flowering near snow melt from 2000 to 2800 meters. It is also known by a later-given synonym *C. biliottii*.

In its best forms it is one of the most beautiful crocuses although the true species is still quite rare in cultivation and under that name fairly often is distributed a sterile hybrid easily distinguishable by the corm tunics and by the very pale color which has nothing in common with the gorgeous bright color of the true *Crocus aerius*. The species name was applied to a range of plants in cultivation, mostly to *C. biflorus* subsp. *pulchricolor*, but the former can be easily separated from the latter by the absence of distinct rings at the tunic basal part. In its best forms it is lighter or darker blue with more or less distinct veining and with a flawless flower form. My favorite samples all are bright blue with a lighter or darker yellow, sometimes even orange, throat surrounded by a wide white zone which at times looks like white rays shining from the deep-colored throat up to half the length of the flower segments. There are deeper lilac forms, too, with darker veining on sepal outsides, and I have seen pictures of almost white blooms with light violet stripes over the back of the sepals. The flower tube is white or has gray or lilac stripes in the upper half. The filaments and anthers are bright yellow. The brightest orange-red stigmatic branches end at the same level as the tips of anthers or rarely, as in my samples, exceed them a little, although Brian Mathew notes that in most cases they end lower.

Crocus aerius is a good increaser and sets seed well, although I cannot list it among the easiest species in cultivation. I think the problem lies in the very specific growing conditions in the wild and in the plant's need for plenty of water in spring and a moderately dry and warm rest period. It is a plant of higher elevations, and in the wild its neighbors are *C. vallicola* and *C. scharojanii*. I have not had success growing *C. aerius* in the open. It is a good grower in pots and blooms abundantly, but in summer I take the pots out of the greenhouse.

Crocus adanensis T. Baytop & B. Mathew

PLATE 140

The phylogenetic tree shows that *Crocus adanensis* (2*n* = 14) is closely allied to *C. paschei*. Fortunately, it is easily separable by the white flower throat. It is named after the province of Adana in southern Turkey from where it is described. It grows on clearings or margins of mixed forests from 750 to 1500 meters and is known from very few localities.

The flowers are light lilac without any blue shading and with somewhat pointed tepals. The outside of the flowers is somewhat buff-toned with a grayish flecked base. The flower throat is white or rarely slightly creamy-colored and this well separates it from other Turkish representatives in the *Crocus biflorus* group. The flower tube on the upper part is grayish speckled. The filaments are whitish, the anthers light yellow, and according to Brian Mathew, sometimes with grayish connectives but I don't have such forms in my collection. The style branches are orange or bright red, exceeding the anthers. The corm tunics are somewhat intermediate; they are membranous and split parallel towards the base but have poorly developed basal rings which sometimes can be absent. For this reason this species is included in the key twice—among species with ringed basal tunics and among those without.

Crocus adanensis is little known in cultivation. In the wild it grows in an area with mild winters and dry summers. I have not yet tried it outside and once almost lost it because of a frost in my greenhouse, but I like it very much because of the shape of its blooms combined with a large starry white throat. I grow it in pots in which it nicely blooms but I have never collected its seeds. Just recently I got another stock and hopefully reciprocal pollination will help to reap the first seed crop.

Crocus tauricus (Trautvetter) Puring

PLATES 15, 141–143

Crocus tauricus (2*n* = ?) is a very special plant for me and perhaps I will be criticized for this name. Brian Mathew in his monograph didn't accept it even at the subspecific level but included the epithet as a synonym of *C. biflorus* subsp. *adamii*. I cannot agree with him here but not because *C. tauricus* was described by Latvian botanist Nikolay Puring.

Taurus is an ancient name for Crimea and is sometimes confused with the Taurus Mountains of southern Turkey. When it comes to crocuses, the situation is even more confusing because both localities have given their names to different crocuses in the same group: *Crocus biflorus* subsp. *tauri* comes from Turkey whereas *C. tauricus* originates in Crimea. Another epithet, *C. tauricus* Steven ex Nyman, belongs to the synonyms of *C. speciosus* and was applied to plants of this species from Crimea.

The population of *Crocus tauricus* (the Crimean crocus) is quite isolated from other forms and subspecies of *C. biflorus*. It grows only on yailas in Crimea at 900 to 1400 meters. It also has very distinct basal rings in the tunic; they are very hard and extremely long toothed.

During my crocus studies I often worked with herbarium sheets of native crocuses of the former Soviet Union, at the Botanical Institute in St. Petersburg. On one occasion I purposely covered the labels of all the annulate crocuses, shuffled the sheets, and then correctly picked out *Crocus tauricus* simply by its basal tunics. There was no mistaking it. In *Flora Taurica* (1929) E. W. Wulff clearly indicated that *C. adamii* (now *C. biflorus* subsp. *adamii*) does not have such teeth on the basal rings and had never been found in Crimea. There is one collection of *C. tauricus* in the Herbarium of the Botanical Institute in St. Petersburg from Krasnodar district near Novorossiysk, a port city close to Crimea, but there are no collections of subsp. *adamii* from this same locality. In addition to its distinct basal rings, *C. tauricus* also has a distinct leaf color—a very bluish gray-green shade.

In 1975 my daughters, both rock climbing national champions in their age groups at that time, went to Crimea with their teams for a training camp. I joined the group to climb the mountains. *Crocus tauricus* was one of my first targets. When we arrived, there was still snow on the Ai-Petri yaila, and *C. tauricus* had only started to flower but the day was cloudy and it was not easy to find the dark-colored flower buds. It took a couple of hours before I managed to collect some 10 to 15 corms because I was also trying to gather as many different plants as I could to cover the variability of this species. The buds had variable amounts of blue and purple striping on the sepal backs. Later I even selected and named several forms, but unfortunately rodents destroyed almost my entire stock. From a few that were saved, I slowly rebuilt my collection.

In autumn of 2008 I returned to Crimea to take pictures of wild *Crocus speciosus* for this book, but this time I went to the Tschatir-Dag yaila. I found that gnawing pests like crocuses in their natural habitat as much as in the garden: in many places I saw remnants of crocus tunics on the surface of the ground. Their very long-toothed basal rings clearly indicated the presence of *C. tauricus* at the site. Compared with *C. speciosus*, which grows in deeper grass and closer to protection provided by shrubs and trees, *C. tauricus* grows in more open places with short grassland and less competition from the neighboring plants.

I revisited the Tschatir-Dag yaila in spring of 2009 for one day. I was very lucky to arrive just three days after most of the snow had melted and on a day without rain. It rained both the day before and the day after. The cloud cover was comparatively thin, and the crocus flowers were half open. They were just at their peak of blooming and the entire diversity could be seen. Previously, on Ai-Petri yaila I had collected only the earliest-blooming plants, all of which had a darker or lighter yellow throat. The anthers ranged from pure yellow to yellow with a grayish or even brownish connective. Here on the Tschatir-Dag yaila, the diversity was really amazing. Most plants had purple flowers; white-flowered specimens accounted for approximately five percent of the population, mostly with more or less striped sepal backs. There were very few pure white specimens (Plate 142), and on the opposite end of the color range there were very few almost blackish purple-colored plants in closed bud. I even photographed one specimen with a reddish-brownish purple flower (Plate 143). It was the reddest-toned crocus flower I had ever seen, more red than the reddest selections of

Crocus tommasinianus. In most cases the back of the outer flower segments was more or less deeply purplish striped or purplish feathered, though there were also specimens with uniformly purple backs of sepals, and some in which the purple was edged with a narrow white zone.

The color of flower throats was quite variable. In the best specimens it was very deep yellow, even somewhat orange-shaded with a brownish tinged margin. The brown shade in the throat was more characteristic of purplish plants; in some it was very dark grayish brown. Sometimes I came across flowers with a dark (grayish or brownish) connective on the yellow anthers, although pure yellow anthers prevailed. The style branches in most cases slightly overtopped the tips of the anthers, but in some they were significantly longer. In a few specimens I observed deep brown stripes on the upper part of the style below the deep orange stigmatic branches.

Crocus tauricus is completely hardy and can be grown outside. It readily sets seed in the garden and in pots. In recent years I grow it under cover but only to protect my small stock from rodents until it increases to more significant numbers.

Crocus biflorus Group

The *Crocus biflorus* complex is the most complicated group in series *Biflori*. At present, 23 (!) subspecies of this variable species have been described. One of them—subsp. *melantherus*—is autumn blooming and was treated (in the proper chapter) as *C. melantherus*, although it blooms through winter till spring. Oh, how much easier it would have been to write this book ten years earlier when there were only fourteen subspecies. I fear that the number will continue to rise.

Every time you get off the beaten track in Turkey you find a different crocus from the *Crocus biflorus* alliance. Almost every population of *C. biflorus* is subtly different from others, and the number of populations is endless. Plants in some populations are very uniform in appearance; in others almost each individual is different. Populations of subsp. *pulchricolor* that I saw at Lake Abant were generally much paler in color and quite uniform compared with the much darker and very variably colored plants from Ulu-Dag.

Knowledge of this group expands with every year thanks to the efforts of two fanatic crocus aficionados from Germany—Helmut Kerndorff and Erich Pasche. The problem is that the majority of the new taxa and several of the older subspecies are still unknown or are very rarely cultivated in gardens. Consequently, there is very little known about their variability, which makes the prospect of compiling an adequate key for their identification very difficult. In this group of crocuses, it is very important to know the origin of the sample to avoid possible identification mistakes.

I have not seen or grown several of the new subspecies (and some older ones as well), though I recently received many of them from both German scientists. My observations of these are too incomplete, so I have relied on the descriptions and pictures of them from other publications. Some of the new subspecies are distinguished by a complex of features which are not easy to include in the keys, so they are included more than once in the proposed key. It would require a magnifying glass to correctly identify taxa for which the presence or absence and the number of ribs in the grooves on the underside of leaves is important. I cannot guarantee that the following key is perfect, but I tried to do my best within the limits of my current knowledge and I have tested the key against samples growing in my collection. Please keep in mind that the variability of this group is extremely wide, so it is very probable that some samples will not coincide with the proposed key.

1. Throat of perianth yellow with white zone surrounding it
    3. White zone prominent, leaves glabrous, flowers striped on exterior .... subsp. *albocoronatus*
    3. White zone less prominent, leaves comparatively hairy, flowers without stripes on exterior .... subsp. *munzurense*
1. Throat of perianth yellow or yellow with dark blotches
    4. Leaves with no prominent ribs in the grooves on lower surface; flowers white or lilac with or without striking purple or brownish stripes on exterior
        5. Anthers yellow, leaves 3–5, throat glabrous
            6. Corm tunics coriaceous, flowers usually striped on exterior, leaves 0.5–2 mm wide .... subsp. *biflorus*
            6. Corm tunics membranous, flowers usually unstriped, leaves up to 1 mm wide .... subsp. *pulchricolor*
            6. Corm tunics splitting longitudinally and with distinct basal rings, flowers without distinct stripes on exterior .... subsp. *caelestis*
        5. Anthers blackish maroon, sometimes yellow, leaves 5–8, flowers distinctly striped, throat yellow, papillose, filaments yellow .... subsp. *stridii*
        5. Anthers black, leaves (2–)4–5(–7), flowers feathered, veined or speckled, throat with bronze-brown blotches, scabrid, filaments brown-violet .... subsp. *caricus*
    4. Leaves with at least one prominent rib in each groove on the underside
        7. Anthers blackish or at least with a blackish connective
            8. Throat yellow with bronze-brown blotches or black staining at edge
                9. Filaments violet only at base .... subsp. *ionopharynx*
                9. Filaments brown-violet throughout .... subsp. *caricus*
            8. Throat yellow without dark blotches, filaments yellow
                10. Leaves 2–3(–4), 1.5–2 mm wide, anthers black .... subsp. *crewei*
                10. Leaves 4–8, 0.5–1 mm wide
                    11. Anthers black with long basal lobes, style branches densely papillose .... subsp. *nubigena*
                    11. Anthers with short basal lobes, style branches usually glabrous
                        12. Anthers blackish before dehiscence, filaments 3–4 mm long, flowers clove-scented .... subsp. *pseudonubigena*
                        12. Anthers with a grayish connective, filaments 4–7 mm long, flowers with sweet honey scent .... subsp. *isauricus*
        7. Anthers yellow

13. Ripe seeds black .... subsp. *atrospermus*
13. Ripe seeds of other color, not black
  14. Stigmatic branches white .... subsp. *leucostylosus*
  14. Stigmatic branches colored
    15. Corm tunic membranous
      16. Leaves 1.5–3.5 mm wide, bracts silvery membranous, flowers pale to mid-lilac .... subsp. *tauri*
      16. Leaves 1 mm or less wide
        17. Bracts, when drying, brownish or reddish, flowers blue-violet .... subsp. *pulchricolor*
        17. Bracts silvery white, flowers finely speckled on the exterior of the outer segments .... subsp. *punctatus*
    15. Corm tunic coriaceous
      18. Flowers finely speckled on the exterior of the outer segments
        19. Anthers yellow
          20. Filaments 4–7 mm long, anthers 7–11 mm long... subsp. *isauricus*
          20. Filaments 2–4 mm long, anthers 11–13 mm long .... subsp. *punctatus*
        19. Anthers with a black connective ....subsp. *isauricus*
      18. Flowers striped on the exterior of the outer segments, if ground color blue—stripes can be less prominent
        21. Sheathing leaves and bracts, when drying, brownish, central stripe on perianth outside very wide, lateral stripes shorter and less prominent, ground color whitish .... subsp. *artvinensis*
        21. Sheathing leaves and bracts silvery white, ground color whitish or bluish to blue
          22. Leaves 3 or 4, filaments flattened, 2–4 mm long .... subsp. *adamii*
          22. Leaves 4–7, filaments filiform, 4–7 mm long ...... subsp. *isauricus*
    15. Corm tunics coriaceous but splitting longitudinally and with distinct basal rings, stigmatic branches deep red, very long .... subsp. *yataganensis*
    15. Corm tunics parallelly fibrous and with distinct basal rings, stigmatic branches orange, equal or overtopping anthers ...... subsp. *fibroannulatus*

Crocus biflorus Miller subsp. *biflorus*

PLATES 144–146

The type subspecies *Crocus biflorus* subsp. *biflorus* ($2n = 8, 10$) is the westernmost subspecies, mainly distributed in Italy but with some populations on Rhodes Island and in northwestern Turkey on both sides of the Dardanelles (provinces of Istanbul and the Çanakkale). It has naturalized in some parts of the United Kingdom, too. The form known as "Scotch Crocus" is a sterile clone of this subspecies though of an unknown origin. In the wild, subsp. *biflorus* grows on limestone formations in open pine woodlands and dry grassy places from 50 to 600 meters. Its name means "two flowers" and refers to its free and abundant flowering.

I only recently obtained wild samples of this subspecies and thus have little experience with their cultivation unless I go back two or three decades when I grew the so-called Scotch crocus and forms known under the names var. *argenteus* and var. *parkinsonii* (Plate 144). All of these were disposed of when I scaled down my collection. I grew these forms only in the open garden where their blooming started on average in the second week of April, always approximately a week later than *Crocus chrysanthus*. All were good increasers but none of them ever set any seed, at least I never put that down in my notebooks of the time. Now I have several corms from Italy originally collected near Palermo.

Flowers of subsp. *biflorus* are basically white with three lilac-blue, purple, or brownish purple stripes on the three outer segment backs and with a yellow throat. Rarely the backs of sepals are only speckled and stripes are absent. The leaves have a narrow median white stripe and lack prominent ribs in grooves on the abaxial surface. In this aspect they are similar to subsp. *pulchricolor* but in the latter the corm tunics are more membranous, not so coriaceous, and the flowers lack stripes on the sepal backs. The bright orange trifid stigma well exceeds the yellow anthers. The leaves usually are equal to the flowers or overtop them during the flowering.

Subsp. *biflorus* is easy to grow both in the garden and in pots. E. A. Bowles described it like this: "Such a hardy, early and free-flowering variety deserves planting in large groups, where its white flowers can open out in the morning sunshine. It has no fads as to soil so long as it is not waterlogged."

Crocus biflorus* subsp. *alexandri (Ničić ex Velenovský) B. Mathew

PLATES 147–149

Two European subspecies of *Crocus biflorus* have white throats. One of these is *C. biflorus* subsp. *alexandri* ($2n = 8$) and my experience with it is similar to my experience with subsp. *biflorus*. Years ago I grew it as a form of a cultivated clone. Then I found that it was very similar to *C. chrysanthus* 'Ladykiller', the only differences between them being the blackish anther lobes and the light blue "tongue" on the base of the outer segments of 'Ladykiller'. I gave preference to 'Ladykiller' as it turned out to be a much better grower and quicker increaser (producing an average of 3.3 corms per year while subsp. *alexandri* produced an average of 2.0). At present I have no samples of subsp. *alexandri* in my collection.

In the wild, subsp. *alexandri* grows in southwestern Bulgaria and in Serbia in mountain scrub or in pastures and stony places on granite and schist at 900 to 1000 meters. It forms a robust plant with large flowers which are white on the inside, but the backs of outer tepals in most cases are deep purple throughout with a narrow white band at the sepal edges. The leaves at flowering time are shorter than the flowers and they have no ribs in underside grooves.

A very unusual form (MO-9528) from Bari in Italy is in John Lonsdale's collection. In bud its flowers look like those of subsp. *alexandri* although it comes from Italy which lies beyond that subspecies's recognized area, but when the flowers open they show a wide grayish blue zone around the white throat. On the outer base of the inner petals is a short blue or even purple "tongue" and the deep purple back of the outer sepals lacks the white band. It seems to be a plant of exceptional beauty, but the question remains—to which subspecies does it belong?

While examining a population of crocuses in central Thrace (Greece) that looked like the usual subsp. *weldenii*, Norman Stevens (United Kingdom) found some flowers in which the back of the outer flower segments was deep purple throughout, the identical color of subsp. *alexandri*. This locality is a long way from the other known populations of both subspecies, the flowers exhibited many variations between these two extremes. Subsequent research showed that plants from this locality have leaves with no ribs in the grooves on the underside and so they belong to subsp. *alexandri*. Ibrahim Sözen found a beautiful albino form growing in European Turkey among plants of typical subsp. *alexandri*. The backs of the sepals were ony slightly speckled.

In the garden subsp. *alexandri* is an easy grower and according to E. A. Bowles, it makes "splendid garden plants, flowering rather late."

Crocus biflorus* subsp. *weldenii (Hoppe & Fürnrohr) K. Richter

PLATES 2, 150 & 151

The other European subspecies with a white throat—*Crocus biflorus* subsp. *weldenii* ($2n = 8$)—is more popular in gardens as it is more frequently offered in bulb catalogs. It is found in northeastern Italy, and along the Adriatic coastal mountains it enters northern Albania. Some of the seedlings I grew from seed reportedly collected in Bulgaria looked identical to subsp. *weldenii*, but I am not sure that they really were wild collected. Subsp. *weldenii* grows in dry grasslands, open stony places, and open woods at 100 to 750 meters.

I have more experience with the cultivated forms that usually appear in various lists than I do with the typical subspecies itself. **'Albus'** has purest white flowers and a white flower tube (Plate 151). In my early years I included it among the best spring-blooming crocuses. It has a very special leaf color that is distinctly bluish. In my collection, only the leaves of *Crocus tauricus* are similar in color, though those are much narrower. Very similar to 'Albus' is a selection distributed under the name **'Miss Vain'** (Plate 2), but it has a slightly grayish flower tube and a very small grayish "tongue" on the outer base of the sepals. **'Fairy'** has bluish flowers and a blue flower tube, and is very similar to *C. chrysanthus* 'Blue Pearl' (differs in the outer base of flower segments). 'Fairy' is very floriferous, producing up to 10 flowers per corm, but how much of the true subsp. *weldenii* is in it remains to be answered. My plants, which I got as authentic subsp. *weldenii*, on the whole mostly resemble 'Albus'. The leaves of subsp. *weldenii* have one or two prominent veins in the grooves on the underside.

In the garden this subspecies is very easy and a good increaser. The plants I grow as wild forms set seed, something I have never observed in 'Albus' or 'Fairy'.

Crocus biflorus subsp. *stridii* (Papanicolaou & Zacharof) B. Mathew

PLATES 152 & 153

Another truly European subspecies, *Crocus biflorus* subsp. *stridii* ($2n = 10$) is not as well known. It was described under the name of *C. stridii* in 1980 but subsequently reduced in status to a subspecies of *C. biflorus*. In the wild it grows in northeastern Greece in sandy, grassy places in scrub or bracken at 30 to 800 meters and seems to be very rare.

It is quite easy to identify this subspecies by the black anthers and the leaves without prominent ribs in the grooves on the underside. The latter feature allows separating it from other black-anthered subspecies of *Crocus biflorus*. Problems can be encountered with yellow-anthered plants which are superficially similar to subsp. *biflorus* in color and in leaf section, but subsp. *stridii* usually has more leaves—five to eight—whereas subsp. *biflorus* only has three to five. In subsp. *biflorus* the anthers are 6 to 11 mm long, but in subsp. *stridii* they usually are longer—11 to 14 mm. My plants come from the mountains near Thessalonica and include both forms. The majority of these have black anthers (Plate 152) and are more beautiful than the yellow-anthered (Plate 153) plants. The basic color of the flower segments is pure white with very contrasting brownish purple stripes along the length of the sepals. The throat is large and very bright orange. The style branches are of the same color and usually shorter than the anthers, but in some specimens they exceed the anthers. The yellow anthers usually correlate in my samples with the somewhat paler striping on the sepal backs and a yellower throat.

My stocks of subsp. *stridii* are still very small and I haven't tried growing them outside nor have I found any reports of anyone else cultivating this subspecies in the open garden. Given its native habitat, it should prefer sandy soil and dry summer conditions.

Crocus biflorus subsp. *adamii* (J. Gay) K. Richter

PLATES 154–156

Crocus biflorus subsp. *adamii* ($2n = 12, 16, 18, 20, 22$—in a wider concept of this subspecies) has a very widespread distribution, reportedly growing from northern Iran through the Caucasus to Bulgaria, southern Yugoslavia, and Turkey in Europe. It is described from Georgia where it grows in various habitats, mostly on calcareous soils, generally in open spots. Not surprisingly given the large area in which it is found, the variability is huge. Zhirair Basmajyan (Armenia) observed populations in which only plants with comparatively small flowers can be found (the Lori region in the western part of Armenia) and others with significantly larger flowers (village of Hartavan near Mount Arai). In the wide sense of subsp. *adamii*, it is found growing at elevations from 250 to 1900 meters.

In my opinion, the Crimean plants of this widely distributed taxon belong to a separate species—*Crocus tauricus*. Furthermore, it may be necessary to create additional taxa in the future, when more is known about the subspecies's geographical variation. At present, I am most familiar with the Caucasian forms of subsp. *adamii*; I recently obtained a few samples from Iran in the eastern end of its range, but still have nothing from the Balkans in the western end of the range.

Just what form in that wide range has to be regarded as typical subsp. *adamii* remains a question because its type locality (a specimen at Kew), according to Brian Mathew, is marked as "Habitat in Tauria et Iberia" (Crimea and Georgia). I regard Georgian plants as typical for subsp. *adamii* because *Flora of USSR* (Komarov 1934–1968) notes that the type specimen comes from the surroundings of Tbilisi in Georgia (the type in Petersburg), and its distribution is limited to the Caucasus and Iran. *Flora Caucasica* (Grossheim 1940) only mentions Tbilisi as the classical location and in the distribution map (No. 228) shows the subspecies occurring only in Georgia, Armenia, Azerbaijan, and Iran, not entering the west of Georgia, Abchasia, and the Russian Caucasus to the west of Ossetia.

In "Wild Plants of Stavropol District of Russia" (Pilipenko et al. 1979) is mentioned a dark blue-violet crocus with feathered backs of the flower segments from the *Crocus biflorus* group under the name *C. artvinensis*. This plant was observed near the famous ski resort of Dombai at an altitude of 2800 meters; however, the lowest altitude limit is given as 2000 meters. The location is on the opposite side of the Caucasus ridge from where *C. autranii* was described. This crocus was collected in the wild in 1971 on the watershed that separates the rivers Maruh and Great Zelencuk, and it bloomed for the first time in the Stavropol Botanic Garden in March 1977. Flowering lasted one week. Leaves appeared after blooming and were green until the end of May. It was reported as a good seeder but later was lost. Is it subsp. *adamii*? The question remains. For *C. adamii* 2800 meters seems far too high an altitude. Of course it is not *C. artvinensis* (subsp. *artvinensis*), which can be found only in Artvin district of northeastern Turkey. Most likely it could be some new, still-unknown-to-science member of the *C. biflorus* group.

Flowers of subsp. *adamii* are very variable. Usually they are light or dark lilac with dark stripes on the outside. Sometimes the flowers are almost white, other samples are lilac with only one darker midvein on the sepal backs. The throat is lighter or darker yellow, in some specimens almost orange-yellow. In darker forms the throat is surrounded by a diffused or starry white zone, sometimes to a certain extent resembling the coloring in the best forms of *Crocus aerius*, but flowers in subsp. *adamii* are much more slender. The style branches usually are shorter than or slightly overtop the anthers. The leaves are shorter than the flowers at blooming time and they are prominently ribbed in the grooves on the underside.

I received a form with smaller, very deep lilac flowers from the Tbilisi Botanic Garden, Georgia, under the name *Crocus geghartii*, but it is now included in subsp. *adamii*. I didn't succeed in finding its original description, only what was written in the *Flora of Georgia* (Ketzkhoveli and Gagnidze 1971–2001) in the Georgian language, which even in orthography is very different from the Latin alphabet. According to Brian Mathew, *C. geghartii* is white or pale lilac with a pale yellow throat, so very different from the plant I got under this name.

In the 1980s I collected another form of subsp. *adamii* near Lake Cherepash'ye at the edge of the city of Tbilisi. The miniature flowers were a variable paler or darker lilac color and had three prominent deep purple stripes on the backs of the sepals. Because of the flower size and color, which were so very different from all other forms of subsp. *adamii* known to me at that time, I followed Georgian botanists who consider it a separate species, *Crocus geghartii*. Unfortunately, all of my stock became rodent food about 10 years later. Not a single corm survived. When I returned to Georgia 25 years later to look for this plant, I found nothing on the site near the lake. I have not found it in other localities suitable for crocuses, so it remains on my wish list.

E. Gabrielian reported about samples of subsp. *adamii* blooming in Armenia in autumn (cited by Takhtajan 2006). Zhirair Basmajyan informed me that individual specimens blooming in autumn in large populations of typical subsp. *adamii* could be found from time to time and it is quite common especially if autumn is favorable. It is not a permanent feature, as the same samples the following season bloomed in their usual time—spring.

Another form of *Crocus biflorus* subsp. *adamii* was collected more to the south by Arnis Seisums. It was part of a very beautiful large-flowering population, extremely variable in color, growing on the Bitschenag pass in Nakhichevan, Azerbaijan, near the Armenian border. The population is the best one I've seen of this very variable subspecies, and the form of the flower is far better than in other forms usually offered under this name. The flowers are blue or violet, tinted or striped darker, very large, and very different from those I had collected near Lake Cherepash'ye. From this stock Arnis selected an abundantly flowering clone with big sky-blue flowers conspicuously striped violet-purple on the outside and named it **'Bitschenag Tiger'**.

Today I have several forms of subsp. *adamii* from western Armenia, where two types of flowers dominate—one with very pale, though prominently striped sepal backs, the other a much deeper lilac with very inconspicuous striping on the sepal backs. I still don't have any sample from the Balkans to compare with the Caucasian plants.

Caucasian *Crocus biflorus* subsp. *adamii* is very easy to grow and has no special requirements either in the garden or in pots. I move my pot-grown plants outdoors during the hottest part of summer as they do not need a very hot, dry summer rest.

Crocus biflorus* subsp. *artvinensis (J. Philippov) B. Mathew

PLATES 157 & 158

In flower color *Crocus biflorus* subsp. *artvinensis* ($2n$ = ?) is very similar to some forms of subsp. *adamii*. It was described from the Artvin district (as *C. artvinensis*), at that time a Russian territory, but now belonging to Turkey. It grows on rocky places and in scrub.

According to the original description, the distinct characteristic of this subspecies is a single central violet stripe on the backs of outer segments. In *Flora Caucasica* (Grossheim 1940) the flowers are characterized as unstriped, and in *Flora of the USSR* (Komarov 1935) as having feathered stripes on the backs of the flower segments. In my plants there are three prominent, closely spaced stripes, so I was uncertain about the correct name. Fortunately, there are some other special characters, which allow separating this subspecies from similarly colored forms of subsp. *adamii* and the yellow-anthered variants of subsp. *isauricus*. It is easily distinguished from subsp. *adamii* by the number of leaves—three or four in subsp. *adamii*, five to eight in subsp. *artvinensis*. Its bract and bracteole are brownish unlike the silvery white bract and bracteole in other Caucasian and Turkish subspecies, and the cataphylls in subsp. *artvinensis* are markedly striated with raised nerves. The yellow stigmatic branches in my plants well exceed the yellow anthers and the leaves can be described as more arched than erect as noted in Brian Mathew's monograph.

Crocus biflorus subsp. *artvinensis* is little known in cultivation and is grown in few collections. I have not found it in the wild. My plants are from corms originally collected by Helmut Kerndorff and Erich Pasche (HKEP-9539). I have tried it only in pots because of my small stock, but judging by conditions in its native habitat, it must readily grow in the open garden, too.

Crocus biflorus* subsp. *isauricus (Siehe ex Bowles) B. Mathew

PLATES 159–162

Crocus biflorus subsp. *isauricus* ($2n$ = 8, 10, 12) is the third subspecies with distinctly striped flowers in most forms. The flowers show the same type of variation as in subsp. *adamii*. The range of distribution stretches in Turkey through the Taurus Mountains in western Antalya province eastward to the mountains north of Silifke(?). The plants grow on vernally very wet open limestone hillsides and flower right after snow melt at elevations from 850 to 2000 meters.

The background color of this very variable subspecies ranges from white, in some specimens even creamy, to lilac, and the sepal outsides can be variously striped or, as in some forms, nicely speckled with purple or gray. Both striped and speckled types can be found in the same population, but striped forms usually dominate. In the best forms the anthers have black or grayish connectives, but there are plenty with pure yellow anthers as well. The corm tunics are very hard, coriaceous.

Forms of subsp. *isauricus* with speckled sepal backs and pure yellow anthers resemble similarly colored forms of subsp. *punctatus*. In this case filament-to-anther proportions within stamens can help separate the two. Subsp. *punctatus* has very short filaments (only 2 to 4 mm long) compared to its anthers (11 to 13 mm long), whereas in subsp. *isauricus* the filaments (4 to 7 mm long) are only slightly shorter than the anthers (7 to 11 mm). The anther lobes in subsp. *punctatus* usually are blackish, but I have noted this only in forms with a blue background color.

The forms of subsp. *isauricus* in my collection are from many localities and show that several key flower types dominate in some of these localities. Along the road from Anamur to Ermenek I found a population with flowers generally white or creamy, even slightly yellow-shaded base color. To the east from Akseki, on the Aldurbe yaila, the flowers were exclusively a uniform bright light blue and the backs of the flower segments were mostly striped, but a few were speckled. Near the next road, only a few kilometers north of Akseki, the variability was far greater. There were plants varyingly striped and speckled and some with almost uniformly deep purple backs of the outer flower segments. In some specimens the throat was so bright orange that it seemed reddish. But everywhere the minority formed plants with black or even grayish connectives on the anthers.

I found the best forms of subsp. *isauricus* on the gentle slopes of the Gembos yaila where they grew alongside *Crocus chrysanthus*, differing only slightly in flowering time. Subsp. *isauricus* dominated in the lower elevations, but higher up, where the snow had melted a few days later, *C. chrysanthus* dominated. As spring progressed, the proportions changed. Had someone filmed this scene in slow motion, we could have seen how the slopes changed color from yellow to bluish. Both species hybridize at this site, producing an array of amazing color combinations, not all of them beautiful. Some are of somewhat dirty shades as if the painter had used colors not in the best proportions when making this rainbow mix, while others are exceptionally beautiful (Plates 162 & 307).

The best of my findings was a brownish tinged, deep orange form of *Crocus chrysanthus*. Very exceptional was a bicolored form of *C. biflorus* subsp. *isauricus*—the backs of the outer tepals were brownish yellow and somewhat sparsely grayish stippled all over, and the inner petals were a shining bright violet; all of this was enhanced by a deep orange throat with a narrow diffused yellow rim and by anthers with distinctly black connectives. Most unusual, and never noticed by me before, was the combination of yellow and purple in the whorls of the flower segments (Plate 162). Also growing at the site were pure white forms as well as nicely striped, speckled, and edged plants with pure yellow or blackish anthers. In fact, all possible colors of crocuses could be seen if you happened to be there at the right time when the two species were blooming in approximately equal proportions. Many other bulbs were also in bloom besides these two crocuses.

Crocus biflorus subsp. *isauricus* is an easy plant in cultivation, possibly not as suitable for the outside garden as subsp. *adamii*, subsp. *tauri*, or subsp. *pulchricolor* because it needs drier conditions during the summer months. Nonetheless, it is very easy in pots. It readily sets seed and has an excellent germination rate.

Crocus biflorus* subsp. *punctatus B. Mathew

PLATES 163 & 164

Crocus biflorus subsp. *punctatus* ($2n = 8$) in some forms can be quite similar to subsp. *isauricus*. It is known from a limited area in southern Turkey where it grows on open grassy places or sparse oak scrub on limestone at 1000 to 1100 meters.

The main difference from other subspecies is in the invariably grayish violet, minutely speckled (covered with small dots) outer surface of the flower segments, a feature from which its epithet is derived. Sometimes this subspecies can be confused with similarly colored forms that rarely can be found among subsp. *isauricus*; in these cases the length of filaments and anthers can help in correct naming (see above). The base color of flower segments can be white to mid-lilac with a deep yellow throat.

The color on the outside base of the sepals at times is somewhat buff shaded. The anthers are yellow with usually grayish toned basal lobes, but this feature can be absent in some forms with a white base color of segments.

Crocus biflorus subsp. *punctatus* is still very rare in cultivation. I grow both color forms only in pots, but my stocks are still too small to be tried outside. In its natural habitat this subspecies experiences hot, dry summers; therefore I keep my potted plants in the greenhouse for the entire summer.

Crocus biflorus* subsp. *crewei (Hooker f.) B. Mathew

PLATE 165

Plants named "*Crocus biflorus* subsp. *crewei*" (or "*C. crewei*") are quite common in the trade, but in fact most plants distributed under this name are the autumn-blooming *C. melantherus*. The true subsp. *crewei* ($2n = 10$) occurs in western Turkey and on the Cyclades growing on rocky slopes in sparse grass or open pine forests from 1200 to 1700 meters.

Most plants offered for sale as *Crocus biflorus* subsp. *crewei* have black anthers, like the true subspecies. The true subsp. *crewei* has only two or three, rarely four leaves, which are wider than those of its close ally subsp. *nubigena*, which normally has more leaves (four to eight). The base color of the flower segments usually is white or rarely light lilac, with three deep purple stripes over the sepal backs. The throat is bright yellow and contrasts well with the long black anthers. The stigmatic branches are deep yellow and they considerably surpass the anthers; the filaments are very short. On the Honaz-Dag in Turkey are populations where typical twin-leafed plants with large jet black anthers are mixed in equal proportions with yellow-anthered specimens.

Only recently I received a few corms of the true subspecies. I grow them in pots as I presume that Latvian conditions would be too harsh to grow them outside. Plants of the true subspecies are very rare in cultivation.

Crocus biflorus* subsp. *nubigena (Herbert) B. Mathew

PLATES 166 & 167

Crocus biflorus subsp. *nubigena* ($2n = 12$) is better known than subsp. *crewei*—most possibly because there have not been any misleading names applied to it. Its name means "born in clouds" and refers to the mountaintop habitat. The subspecies grows in rocky places or sparse scrub, clearly preferring granite or schist at 650 to 1850 meters. It is found in western and southwestern Turkey and on some of the Aegean Islands.

Subspecies *nubigena* is most closely allied to subsp. *crewei*, but differs in having more leaves. Flower color is quite variable, from white to light violet with dark striped backs on the outer flower segments. The throat color varies from lighter to very deep yellow, in darker forms even with a somewhat orange shade, and rarely with grayish flush over the margin zone of the throat. Flowers open widely in the sun displaying the large black anthers. Their basal lobes are very long and slender. The style is yellow to light orange, more or less on the same level with the anther tips. Subsp. *nubigena* was originally described by William Herbert who characterized it thus: "Of three bulbs which survived . . . one had white flower speckled without on the sepals, and two had the flowers suffused and feathered with very dark brownish purple, one of them having the anthers nearly black on the outside, though the pollen was yellow, the other having them yellow." I have never seen yellow-anthered plants in subsp. *nubigena*, but such can be found in "black-anthered" subsp. *pseudonubigena*.

Crocus biflorus subsp. *nubigena* is a very floriferous subspecies. Three large corms in a 10-cm-wide pot can cover the entire surface of the pot with flowers. The subspecies readily sets seed which has a good rate of germination. Regardless of my large stocks and the good rate of increase, I still grow this subspecies only in the greenhouse, in pots and on a bed, but haven't tried it outside.

Crocus biflorus subsp. *pseudonubigena* B. Mathew

PLATES 168 & 169

Although *Crocus biflorus* subsp. *pseudonubigena* ($2n$ = 18, 20) in some forms is similar to subsp. *nubigena*, it is well-separated geographically, being a plant of rather dry regions in southeastern Turkey, where it grows in dry oak maquis at 500 to 1300 meters.

The flowers of subsp. *pseudonubigena* have comparatively narrow (4 to 7 mm) segments and in general appearance are slenderer than those of subsp. *nubigena*. This trait, along with short filaments and short basal lobes of the anthers, makes identification of the subspecies quite easy. The flowers have the nice scent of carnations and they are quite variable in color. In my samples the base color of segments is white, but it can also be lilac. All my specimens of subsp. *nubigena* are variably striped on the back of sepals, but the plants of subsp. *pseudonubigena* are more variable—from striped to almost pure white with a narrow purplish "tongue" on the outer base of sepals. The filaments are very short and anthers distinctly black prior to dehiscence although in several specimens from the same locality (Hatkis-Dag) they are yellow. The throat is bright yellow. The yellow style branches end well below the tips of anthers. Because the subspecies is known from very few collections, I believe its variability can be much greater. All the samples known to me have come from the same plants originally collected near Gaziantep in a sparse oak (*Quercus coccifera*) scrub on limestone formations at an elevation of 1100 meters by KPPZ expedition in 1990.

Crocus biflorus subsp. *pseudonubigena* is a plant from very hot and dry summer conditions so it is suitable only for growing in pots in my climate. I leave it in the greenhouse under cover during the summer. It readily sets seed but is quite slow to increase by division.

Crocus biflorus **subsp.** ***ionopharynx*** Kerndorff & Pasche

In 2004, Helmut Kerndorff and Erich Pasche recognized two more subspecies with black anthers—subsp. *caricus* and subsp. *ionopharynx*. The following information refers to the original description.

Crocus biflorus subsp. *ionopharynx* (meaning "violet-throated") is somewhat similar to subsp. *pulchricolor* in its usually uniformly violet-blue flowers, but is easily distinguishable from the latter by the very long filaments (equal in length to the anthers) which change from yellow at the apex to violet at the base. *Crocus pestalozzae* has a similar dark-stained filament base, but its flowers are mostly white or rarely pale blue. The throat of subsp. *ionopharynx* is dark orange-yellow with black staining at the top of the yellow zone and a violet zone at the bottom of the throat, but this is visible only in a dissected flower. Kerndorff and Pasche observed distinct striping or feathering on the outside only in some specimens. The style ends at the tips of the anthers or is longer. The corm tunics are hard, coriaceous. The subspecies was found only on non-calciferous soils on Bati Menteşi mountains (province Muğla) at an elevation of 800 to 1200 meters and is allegedly "not easy in cultivation."

Crocus biflorus **subsp.** ***caricus*** Kerndorff & Pasche

PLATE 170

The second black-anthered subspecies described by Kerndorff and Pasche in 2004 is *Crocus biflorus* subsp. *caricus* (Plate 170). The flower is basically white with purple stripes on the backs of the flower segments. The subspecies is known from several localities in Caria, southwestern Turkey, where it is locally abundant at elevations of 900 to 1200 meters. It is somewhat similar to subsp. *crewei*, but differs in more numerous, though smaller leaves and distinct bronze-brown blotches in the yellow throat. In some forms the throat at the base is dark red-brown becoming orange and then yellow at the top. The filaments are very long and brown-violet throughout—a unique feature in crocuses—so it will not be difficult to identify this subspecies.

In the wild *Crocus biflorus* subsp. *caricus* grows on both calciferous and non-calciferous formations. It can be found on margins of pine forests, sometimes among or under *Castanea sativa*. Our team found it side by side with autumn-blooming *C. nerimaniae*.

A plant with similarly colored throat and filaments but with blue flowers and with yellow anthers (Plate 171) was discovered by our team near Malatya (BATM-385) in Turkey, far to the east from subsp. *caricus* localities. Most possibly it is new subspecies but more research in the wild is necessary.

Two subspecies from northern Turkey have membranous tunics and are the easiest to cultivate of all the Turkish subspecies of the *Crocus biflorus* complex. They occur wild in opposite parts of Turkey—subsp. *pulchricolor* in the west, subsp. *tauri* in the east.

Crocus biflorus **subsp.** ***pulchricolor*** (Herbert) B. Mathew

PLATES 172 & 173

Crocus biflorus subsp. *pulchricolor* ($2n = 8$) is very common among crocus lovers and in the wild is distributed in northwestern Turkey where it grows in damp alpine meadows, especially in the Ulu-Dag and near Bolu where it flowers by the melting snow on edges and in clearings in pine woodlands from 1000 to 2300 meters. It is one of the plants to which the name "*C. aerius*" is sometimes wrongly applied.

The flowers of subsp. *pulchricolor* are invariably without stripes on the sepal backs and their color can vary from very light violet to deep violet and even purple in some specimens. Instead of stripes there is a typical dark "tongue" at the base of the outer flower segments which sometimes stretches quite high up the middle of the sepal or even a little higher. Quite often there are grayish basal blotches surrounded by a darker blue or purple zone extending up like one or two rays. Very rarely it reaches the sepal tops in a form of a dark median zone. The throat is large, deep yellow, the filaments and anthers are yellow. The style branches are bright orange terminating at the tips of the anthers. The leaves are very narrow—only up to 1 mm wide.

I saw the greatest variation in flower color on the Ulu-Dag where this subspecies grew side by side with *Crocus herbertii* and in some places together with *C. chrysanthus*. In overall appearance some of the plants clearly looked like hybrids with *C. chrysanthus*. Another population near the road from Bolu to Nallihan was more uniform in color, in general much lighter than at the Ulu-Dag. Some plants were very light, but none was pure white. In both spots they grew on very wet soil, where the feet sank while walking.

My first stock of subsp. *pulchricolor* came from Chris Brickell in the early 1990s. This stock (BM-8514) seems to be the most vigorous and widespread in the gardens, and is now offered by several nurseries. It is very uniform and perfectly reproduces itself from seed, showing minor if at all variation in flower color, which is clearly bright lilac-blue on the outer tepals and somewhat lighter on the inner. The color fades as the flower ages and at the end the inside of the inner tepals becomes almost white. The throat is large, bright yellow.

Crocus biflorus subsp. *pulchricolor* is a perfect grower in the open garden as well as in pots, and until the winter of 2005–06 I regarded it as completely hardy. That winter most of crocuses seriously suffered or even died. Its relative from the eastern part of Turkey—subsp. *tauri*—turned out to be much tougher and, growing side by side, suffered considerably less. *Crocus biflorus* subsp. *pulchricolor* readily sets seed and even self-seeds on paths between the beds in my nursery.

Crocus biflorus **subsp.** ***tauri*** (Maw) B. Mathew

PLATES 174–176

Crocus biflorus subsp. *tauri* ($2n = 20, 22$) is the second subspecies from northern Turkey having a membranous tunic. It is somewhat similar to subsp. *pulchricolor*, but grows in the opposite part of Turkey, occupying a very large area from central Turkey to northwestern Iran and northern Iraq, where it grows on rocky slopes, in mountain steppe and sparse oak scrub at higher elevations. It blooms by melting snow at elevations from 1330 to 2800 meters sometimes on the most claggy soil imaginable. I encountered its flowers in northeastern Turkey even in the first half of June.

In flower color subsp. *tauri* is very similar to subsp. *pulchricolor*, but it has wider leaves. The corm tunics are somewhat thicker, but still membranous. The flowers are lighter or darker lilac-blue without stripes on the backs of segments, though sometimes they are finely feathered. The throat color has to be light to deep yellow, rarely even slightly orange-shaded in the depth of the throat.

Crocus biflorus subsp. *tauri* grows freely both in the garden and in pots, but the suitability can vary depending on the origin of the sample. Plants from the northern part of the range are easier in the open garden and in my garden turned out to be hardier than plants of subsp. *pulchricolor*. Plants from the surroundings of Bingol further south are less hardy. Perhaps it would be safer to grow the plants from the southern and eastern part of the area in pots.

In Iran, where *Crocus biflorus* is represented by only subsp. *tauri* and subsp. *adamii*, I found a population with very pale, almost whitish flowers (Plate 176). The backs of the outer segments of some specimens had darker purplish feathering, but the throat color varied from very pale yellow to dark yellow. Like the flower color, the stigmatic branches were also very pale yellow. The wide leaves excluded subsp. *adamii*, and the silvery bract and bracteole along with the leaf characters made me think of subsp. *tauri*, but I found this population far east of Tehran, between Astaneh and Fulad Mahalleh on open wide slope at an elevation of 2350 meters. The site is approximately 700 to 800 kilometers east of the easternmost locality of subsp. *tauri* known before. The crocuses grew in a shallow gully and many were under running water with flowers rising above water. The shape of the flower segments was narrower and more pointed than any other samples of subsp. *tauri* and subsp. *adamii* that I had seen. Possibly the population represents another undescribed new subspecies of *C. biflorus*. Oh, when will this finally end? I am more and more inclined to agree with Brian Mathew and regard all forms as belonging to one extremely variable *C. biflorus*.

In recent years several new subspecies of *Crocus biflorus* were discovered with yellow anthers. Several of them have some very distinct features which allow easy separation from the others. All of them were discovered and described by Helmut Kerndorff and Erich Pasche and I express my greatest thanks to both for sharing the living material with me to test in the conditions of Latvia. The following information about these crocuses comes from the official publications of these scientists.

Crocus biflorus* subsp. *atrospermus Kerndorff & Pasche

PLATE 177

Crocus biflorus subsp. *atrospermus* is unique among crocuses with its seeds which turn black when dried. This feature is referenced in scientific name, which means "black seed." In other aspects this subspecies is intermediate between subsp. *tauri* and subsp. *pulchricolor*; it is very colorful and lacks stripes on the backs of flower segments. It differs, however, in having coriaceous corm tunics, whereas its relatives have membranous tunics. Subspecies *atrospermus* is leafier and usually has seven to nine leaves, while the other two subspecies have no more than five leaves. Very prominent is the dark yellow throat visible from the outside, and the presence of a dark violet spot or "band" on the outside base which is particularly prominent on the outer segments. This subspecies is known from several sites in the Boncuk mountain range in southwestern Turkey.

I once collected a few corms on a spot (at an elevation of 1480 meters) very close to the place from where the type specimen comes. It was a very frosty morning—all springs and streams from melting snow were stiffened by the night frost and had turned into beautiful "ice falls." Later in the day it began to snow and I was forced to return to the hotel. I found the crocus flowers, almost killed by the frost, between stones on the tops of very steep slopes. In color the flowers perfectly matched subsp. *atrospermus* and afterwards it turned out that the leaf characteristics corresponded as well. Later I found that several other acquisitions which I had collected from this district belong to subsp. *atrospermus*. The problem is that to be certain in verification, the seed crop is needed. The corms were easy to cultivate, but as with other rarities I didn't try them in the outside garden.

Crocus biflorus* subsp. *leucostylosus Kerndorff & Pasche

PLATE 178

Crocus biflorus subsp. *leucostylosus* is very special with its pure white erect stigma and was named for this characteristic. The subspecies is still known only from the type locality, although there are records about occasional plants with white stigmas among other species. In the *locus classicus* all the plants invariably have white styles. Subsp. *leucostylosus* is one of the few subspecies with a membranous corm tunic. The leaves are very narrow, the anthers are pure yellow. The subspecies is described from the province of Denizli where it grows in alpine grassland, damp meadows, and soaked borders of mountain streams, and is locally abundant. Judging by the growing conditions in the wild, it should be quite easy in cultivation. Kerndorff and Pasche suggest that growing conditions can be similar to those of subsp. *pulchricolor*. In the wild subsp. *leucostylosus* grows near *C. chrysanthus* and occasional hybrids between them have been found. I'm growing it quite well in pots, but my experience still is very short.

Crocus biflorus* subsp. *albocoronatus Kerndorff

The most distinctive feature of *Crocus biflorus* subsp. *albocoronatus* is the white corona surrounding the yellow throat, as the name, which means "white crown," infers. The subspecies was discovered in the central Taurus Mountains and turned out to be a very local plant thinly scattered in rocky outcrops among shrubs and coniferous woodland. Its closest relatives are subsp. *nubigena*, subsp. *pseudonubigena*, and subsp. *artvinensis*. All of these are more leafy, having usually five to eight leaves, but subsp. *albocoronatus* has three or four leaves and they are wider (1.5 to 2 mm comparing with 0.5 to 1 mm in its relatives), thus compensating for being fewer in number. The cataphylls and bracts become intensively brown with the approach of dormancy, as they do in subsp. *artvinensis*. The flowers are large and elongated with relatively narrow segments and a slender overall shape without the "waist" that is typical of most variants of *C. biflorus*. In the sun the flowers open widely, creating a starlike appearance. This subspecies has the largest seeds in the entire *C. biflorus* aggregate.

The growing conditions look quite similar to those of *Crocus abantensis* and *C. kerndorffiorum*. *Crocus biflorus* subsp. *albocoronatus* requires a dry but not hot summer. It is very rare in cultivation as well as in the wild and so the experience of growing it is very limited. I recently received a few corms from Erich Pasche; they grow in a pot that stays outside the greenhouse in summer, moving back inside in September.

Crocus biflorus* subsp. *yataganensis Kerndorff & Pasche

Crocus biflorus subsp. *yataganensis* was found in the Doğu Menteşe Mountains near the city of Yatağan in the province of Muğla, Turkey, where it grows at an elevation of 1000 to 1200 meters on calcareous formations in open forests along the mountain slopes. Its flowers are richly violet with a distinct dark violet blotch towards the perianth tube. The anthers are yellow; the style branches are long and deep red. The corm tunics are coriaceous, splitting longitudinally in numerous stripes, and have basal rings at the base. According to Helmut Kerndorff and Erich Pasche, the comparatively long and deep red style branches of an open flower are most remarkable, and similar ones in the *C. biflorus* complex can be observed only in the autumn-flowering *C. wattiorum*. Subspecies *yataganensis* is somewhat similar to subsp. *pulchricolor* (it has membranous tunics) and subsp. *ionopharynx* (anthers are black). I suppose that it would like dry and warm summer conditions.

Crocus biflorus* subsp. *caelestis Kerndorff & Pasche

PLATE 179

Crocus biflorus subsp. *caelestis* is named for its rather pale "heavenly blue" flowers with very diffused or even absent featherings and markings on the outside of the sepals. The sky-blue color is most intense at the top of the flower segments, but quickly fades to white toward the white or light lemon-yellow, rarely darker throat. The filaments are colorless or light yellow; the anthers are yellow. The style branches are shorter or equal to the anther tips. The corm tunics are more or less membranous, splitting longitudinally in numerous stripes, and with basal rings at the base. This subspecies differs from the allied subsp. *punctatus* in the pale color and in the lack of dark tips on the anther lobes; subsp. *pulchricolor* is much more colorful and has a deep yellow throat. Subspecies *caelestis* was found on calcareous soils near the city of Uschak in open areas and light forests at an elevation of 1200 to 1400 meters. No hybrids have been observed with the neighboring *C. chrysanthus*. Judging from its native habitat, it would like dry and warm summer conditions.

Crocus biflorus **subsp.** ***fibroannulatus*** Kerndorff & Pasche

PLATE 180

Crocus biflorus subsp. *fibroannulatus* was described from northeastern Turkey in the province of Artvin. It has very strange corm tunics—a combination of fibrous and annulate types from which the name is derived. The corm tunics consist of parallel fibers and more or less distinct rings at the base. The subspecies was found on west-facing volcanic slopes with a thin layer of humus. Its flowers are more or less lilac with violet stripes in combination with differently shaped featherings. The stripes are visible on both sides of the outer and inner segments and in this aspect the subspecies somewhat resembles *C. aerius*, but the tepals of the latter usually are wider and more bluish.

In the somewhat similar subsp. *artvinensis* the base color of the flower segments is whiter and the corm tunics look very different. Judging from pictures I've seen of these two subspecies, subsp. *artvinensis* has yellow style branches and a uniformly purple flower tube while subsp. *fibroannulatus* has orange style branches and a purple-striped flower tube.

In the wild subsp. *fibroannulatus* was discovered in quite specific habitats, so I have no idea about the conditions required in cultivation. I suppose that they could be similar to those of the little-known subsp. *artvinensis*.

Crocus biflorus **subsp.** ***munzurense*** nomen nudum

PLATE 181

The name "*Crocus biflorus* subsp. *munzurense*" has not been officially published, but it is used by Helmut Kerndorff and Erich Pasche in their 2003 article about *Crocus biflorus* in Anatolia. The authors note that subsp. *munzurense* seems to be connected with subsp. *pseudonubigena*, but it has odd sky blue to lilac flowers without stripes on the sepal outsides and a white zone above the yellow throat. It also has very hairy leaves.

This subspecies is not well known and the place where it comes from (the Munzur mountain ridge) is not easily accessible at present as it is one of the centers of the Kurdish resistance and entry to the valley is strongly controlled by the Turkish military forces. Everything hinges on the actual situation at the moment. Our team was stopped at the gendarme post before the entering the valley and protractedly questioned why we just wanted to drive through the Munzur valley. In the end we were allowed to proceed after being warned to not stop anywhere or talk to any of the local people. We did stop to have lunch and to do some botanizing by the roadside. It is a pity that such a beautiful place for tourist now looked abandoned. Instead of once-big roadside restaurants and petrol stations, only ruins remained.

A couple of samples—one from the original collection (HKEP-9911), another from the team of Jim Archibald, Norman Stevens, and Arnis Seisums (SASA-211)—seem quite easy to grow in pots.

The Eastern Runners—Trio from Central Asia

This group of closely allied species grows in the easternmost area of the genus's distribution range and consists of the earliest bloomers in spring. In cultivation they start flowering with the melting snow, but in some seasons with unusual weather conditions they can bloom in early winter (December) if grown outside; and in pots and in exceptional cases, even in late autumn (November). All the species are joined in series *Orientales* and all have the same chromosome number—$2n = 20$, although their areas are separated and don't overlap. Regardless of the identical chromosome numbers, I have never succeed in crossing them in any thinkable combination.

According to the new phylogenetic tree, the western neighbor of these eastern species is the autumn-flowering *Crocus caspius*, which is very close to the Central Asian crocuses. Because of the corm tunic shape Brian Mathew placed *C. caspius* in series *Biflori*. All the Central Asian species have one quite distinct feature—they keep the seed capsule at soil level even when the seeds are ripe.

Because there are only three species in this group, no key is needed. Identification in this group is quite easy. If the flowers have lighter or darker blue suffusion on the sepal backs and a bluish throat, or if the corm tunic is fibrous and distinctly netted, the plant is *Crocus michelsonii*. In the other two species the corm tunics are rather membranous, splitting into many parallel fibers sometimes joined towards the apex. Other features separating *C. michelsonii* from the relatives are the absence of any yellow in the throat and the white stigma. *Crocus alatavicus* has white flowers with more or less gray or even blackish purple stippling on the sepal backs, while *C. korolkowii* flowers are basically yellow; white specimens are extremely rare.

Crocus michelsonii B. Fedtschenko

PLATES 182–184

Crocus michelsonii ($2n = 20$) is the westernmost species of the Central Asian crocuses, growing wild throughout the Kopet-Dag mountains bordering Turkmenistan and Iran. It is found on open stony hills and *Artemisia* steppe up to the province of Gorgan (village of Almeh) in the west from 1220 to 2300 meters. Its name commemorates A. Michelson who collected plants in Central Asiatic Russia. Although it is among the most handsome spring-flowering crocuses, it unfortunately is one of the most difficult in cultivation.

Flowers of *Crocus michelsonii* are large; the inside of the segments is white with a larger or smaller bluish or purplish diffused throat zone, which sometimes is very light, even whitish. On the outside the background color is whitish more or less heavily suffused or speckled lilac-blue, at margins usually paler, sometimes the margin is white; the tube normally is lilac. The style is white and equals or slightly exceeds the length of the yellow anthers. A plant produces four to seven, rarely as many as nine rather grayish leaves.

I first encountered this species in the Arvaz Valley on the Turkmenistan side of the Kopet-Dag. I was there at the end of May, so its flowers had faded long ago, but its leaves covered the ground like grass near our camp. It was impossible to take a step without trampling on the crocus leaves. In that setting I for the first time saw seed capsules with almost ripe seeds still staying underground, making it impossible to collect seeds without digging up the corm.

In my first Latvian garden, I grew samples from this population in coarse sand and they proved to be quite vigorous. The plants set seed annually, and increased moderately by divisions (at Arvaz I found clumps of up to five corms, but mostly only one or two). I even made some selections by flower color. I was surprised by the very large seeds, the largest among the crocuses with which I was familiar at that time. In the garden the plants bloomed simultaneously with the melting of snow, sometimes pushing the flower buds through last snow patches. After my garden moved to a new place where the soil was heavy clay, my stock very quickly decreased and within a few years I was left with nothing.

In the meantime Leonid Bondarenko grew this species very successfully. His climatic conditions are more continental and the soil is very light. He has selected several cultivars but because their rate of increase is very moderate, their future is quite doubtful. The most impressive ones are **'Odissey'** (Plate 184) with nice pale lilac, creamy edged sepal outsides. On the opposite side of the spectrum stands **'Turkmenian Night'** with very dark lilac flower segments. In some seasons the plants increase quite fast, producing up to three corms in the place of one, but in other seasons the stocks suffer and I can lose the greatest part of them.

Now I grow all my *Crocus michelsonii* samples only in the greenhouse. Most of them originate from Leonid's collection. Even under cover they are not very easy just because of the early blooming. The earliest recorded flowering in my collection is on 16 November though flower buds in late December are quite usual. The weather then is dull and dark, so the flowers open very rarely. To get seeds I usually take the pots inside where flowers open and I can hand-pollinate them. In recent years the "normal" blooming in the traditional spring time is quite rare.

Some years ago Henrik Zetterlund (Sweden) introduced a later-blooming form (T4Z-1116) from higher elevations on the Iranian side of the Kopet-Dag. Usually the samples of a species collected at higher elevations bloom later than those collected at lower elevations, even when grown in the garden. In my collection, this later-blooming form always blooms only when true spring has started in the greenhouse.

In general, the Iranian forms bloom somewhat later than the Turkmenian ones. I have seen *Crocus michelsonii* in many locations on the Iranian side but nowhere as abundant as in the Arvaz Valley. In Iran I saw it mostly as individual plants more or less densely scattered on the slopes. The variability of the flower color was of the same wide range but the flowers always had their very specific lilac shade on the sepal backs and the lilac throat.

In Latvian conditions *Crocus michelsonii* is difficult in the open garden and not very easy even in pots. It is quite rarely offered by bulb nurseries. It requires very wet conditions at flowering time and a little after, but as soon as the leaves start to turn yellow all watering must be stopped and a dry and hot summer rest provided until late autumn. In seasons when flowering starts in spring it readily sets seed even without additional hand-pollination.

Crocus alatavicus Regel & Semenov-Tjan-Schansky

PLATES 185 & 186

Crocus alatavicus ($2n = 20$) is somewhat easier to grow in the garden than *C. michelsonii*, but it does not increase as well. Occasionally a corm might produce two or three replacement corms but that doesn't happen regularly and therefore the clone selection is fruitless or very slow. Propagation of this species occurs almost entirely by seed.

In the wild *Crocus alatavicus* is the most eastern-growing crocus species, distributed from the Tashkent area in Uzbekistan through the Dzungarian Ala Tau to the northwestern Xinjiang province in China. It is the only crocus native to China. It grows at high elevations (1800–2300 m) in mountain meadows where it flowers by the melting snow.

Crocus alatavicus is one of the earliest of the true spring-blooming crocuses. My records show that one year it started to flower in the open garden on 13 February and the next year even earlier, on 14 January.

Samples in my collection come from Uzbekistan (mostly from the Chimgan Range and the surrounding ridges), Kyrgyzstan, and Kazakhstan. On the whole, they are more or less identical, but markings on the sepal backs vary greatly; some plants have black stippling, other purplish, and the amount of color varies from a completely covered outside to a narrow middle stripe or two.

Some varieties of this species have been described based on the color of the flower. During one of my trips I searched for the form known as var. *albus* by Eduard Regel. This form has a white back on the flower segment. Finding such a plant was very difficult as the flowers of *Crocus alatavicus* are very photoactive; they open widely with the first rays of sun, showing the invariably white tepal insides. Selection can be done only when the flowers are closed, which happens in rainy weather when walking along the slippery mountain slopes is almost impossible. Very rarely plants without yellow in the throat are found.

On warm days flowers of *Crocus alatavicus* emit a fresh violetlike scent similar to that of *C. michelsonii*. The style is orange or yellow (few records of a white style exist), equal or exceeding the anthers. The plant is leafier than *C. michelsonii*, and flowering-size corms usually produce 8 to 15 leaves, rarely fewer. The leaves are green not grayish as in *C. michelsonii*.

Crocus alatavicus blooms soon after the snow melts. At Chimgan, it blooms at the same time as *Colchicum luteum*. Our team was there in 1996 when one very cold night the temperature outside dropped well below zero. In the morning the inner side of our tent roof was covered with ice, but the day promised to be very sunny. Not a single cloud was in the sky. In the early morning the foliage of nearby bulbous plants looked as if it had been boiled, but surprisingly soon it recovered completely. The bright sun opened the flowers of *Crocus alatavicus* and I could select for my collection the most ornamental forms. Although the plants were in full bloom, the corms still lay in frozen soil and thus were quite brittle. The soil was very wet, even damp, but as the snow receded the soil quickly dried out.

Crocus alatavicus is a very good plant, though its beauty can only be appreciated in full sunshine; because the backs of the sepals are stippled and spotted with ash gray or purplish gray, the closed flowers are almost invisible. Recently I received some plants from the Bishkek Botanic Garden (Kyrgyzstan) collected on Kashka-Su at an elevation of 2200 meters. One of the plants had a single, narrow gray stripe from the base to the tip of the sepals. Another type of variation in *C. alatavicus* is seen in the color of the backs of outer segments. Plants with a yellowish base color were described by G. Baker as var. *ochroleucus*. Such specimens are scattered among plants with a white base color. The shape of the tepals, which can vary greatly from narrow and pointed to wide and rounded, is a useful feature by which selection of more ornamental types (strains) can be made.

A very unusual form of the species was found below Great Chimgan by the late Václav Lajn from Czech Republic. The outer flower segments in this plant were violet both inside and outside, but as is typical of this species, the plant was a very shy increaser. Lajn grew the plant in a frame in a pot since 1987 and during more than 10 years got only eight corms altogether (later lost).

E. A. Bowles reported in 1952 that *Crocus alatavicus* was almost lost in cultivation. In the 1980s I brought 500 corms of it to Michael Hoog in Holland and from his nursery it again found its way into collections. I have been growing this species with varying success: my stocks bulk up, they set seed, and then one season I'm back at the starting point. At this writing I have passed such a period of decline; my stock is beginning to increase from a dozen corms that survived a severe black frost. The plants set seed, but due to the very early blooming time must be pollinated by hand. I have a bee hive in the greenhouse but since the sun rarely shines at the time this crocus is blooming, the bees stay in their hive rather than fly around in search of nectar.

At present, I grow *Crocus alatavicus* only in the greenhouse, despite having successfully grown it in coarse sand in my first garden. At flowering time it needs rather wet conditions, but in summer it prefers dry, warm rest and is less tricky than *C. michelsonii*.

Crocus korolkowii Maw & Regel

PLATES 187–196

The easiest of the Central Asian crocuses in cultivation is *Crocus korolkowii* ($2n = 20$). It is very easily separable from other species by its yellow and very fragrant flowers. It also is the leafiest Central Asian crocus, usually producing 10 to 20 leaves per shoot.

In the wild it is distributed from Uzbekistan, where it grows northwards up to the Kara-Tau Mountains and southwards, through Tajikistan to northern and eastern Afghanistan and northern Pakistan. It occurs on open rocky and grassy places from 1200 to 3150 meters. In Uzbekistan its area very closely approaches the area of *Crocus alatavicus*, but they nowhere overlap. *Crocus korolkowii* is named after General Korolkow who collected specimens from which this species was described.

During one of my first trips to Central Asia on the Aman-Kutan (now Tahta-Karacha) mountain pass I suddenly spotted a white flower among the masses of yellow-blooming *Crocus korolkowii*. It seemed incredible that *C. alatavicus* would be growing alongside *C. korolkowii*. Everywhere in literature it is stated that, although their areas can come into contact, these two species never form mixed populations. Careful checking of the picture proved that the color pattern on the outsides of the flower segments was more typical of *C. korolkowii*. Later discoveries of very rare white-colored plants far from the area of *C. alatavicus* confirmed that albinos can be found even in this species that previously was regarded as having only yellow flowers.

My first encounter with *Crocus korolkowii* was in the upper course of the river Agalik in the west end of the Serawschan mountain ridge not far from Samarqand, Uzbekistan. Entire slopes were almost completely covered with the leaves of this species. The number of plants growing there was enormous. They were everywhere—in the alluvial meadows and in the splits of the rocks. Many of them had a surprisingly high rate of vegetative reproduction: some clumps consisted of as many as 20 corms. Later I found that this population showed the greatest variability in flower color. One plant differed from the next in the brown striping, or in the stippling on the back of the outer flower segments, or in the shape of the tepals. Most of my named *C. korolkowii* varieties have come from here.

Populations in other places that I later visited were more or less uniform in color and the brightest color was at Agalik. Only at the Aman-Kutan pass, situated not very far, the variation could be compared with that of the Agalik valley. Brian Mathew writes that similarly rich variability was recorded by Paul Furse in plants from Afghanistan. Elsewhere, all the samples collected by me were very uniform in color without any prominent design on the sepal backs.

Abundant flowering is quite characteristic of *Crocus korolkowii*. Many clones produce up to 15 to 20 blooms per corm. They are excellent increasers, forming very large corms.

Among the best forms which I named as varieties is **'Lucky Number'**, the earliest of them (Plate 190). Originally it was selected as clone Number 12 and later named by me 'Number Twelve'. It was much liked by the Dutch nurseryman Jan Pennings, only the name didn't appeal to him, and he proposed renaming it 'Lucky Number'. It really proved to be a "lucky number" in the horrible winter of 2005–06, when a very warm January in which all the crocuses were in full bloom was followed by February with a full snowless week of minus 35°C. 'Lucky Number' was one of the few crocuses which survived in my open garden.

'Yellow Tiger' (Plate 193) is another very beautiful variety with large yellow flowers intensively speckled and striped brown along the backside of the sepals. It is widely grown in Holland now as well. **'Mountain Glory'** is very early blooming. It has bright yellow blooms with a wide brown stripe up to the middle on the outside of the sepals. It is one of my favorites, especially when the flowers are half-opened. **'Kiss of Spring'** has very large, rounded, pure deep yellow flowers. **'Dark Throat'** (Plate 187) was named for its dark brown throat, although several of my varieties of *Crocus korolkowii* have brown throats.

As I write this book, the biggest part of Dutch-grown varieties and forms of *Crocus korolkowii* originates from collections I gathered at the upper course of Agalik in 1976. These plants replaced the earlier-grown sterile form which was somewhat similar in outer appearance to 'Dark Throat', but was smaller and a weaker grower and bloomer. The fastest increaser is **'Spring Cocktail'**, which once produced 10 corms from one planted corm after one year of growing but averages 5.2 new corms per planted corm.

In May 1981 our team visited the Varzob gorge in Tajikistan. We set up base camp in Chinoro, a side gorge near the middle of the Varzob. In a split stone I found large clumps of *Crocus korolkowii*. Here it mostly grew in clumps like in Agalik. Hoping to find a similar variability, I collected many groups from different places, but great was my surprise the next spring when they all bloomed with very uniform, large, rounded, bright yellow flowers with grayish green shaded sepal backs.

In 1982 I again collected a few corms at much higher elevations in the Varzob gorge at the village of Hodji-obi-Garm where the species covered the mountain plateau like grass. The plants again were of the same color type as those collected at Chinoro. I selected the best increaser and named it **'Varzob'** (Plate 192), but all the time I thought that it was less spectacular than other clones, at least when the flowers were not open in the sun. Surprisingly, this variety received the Preliminary Commendation award from the Royal Horticultural Society in 1977. Alan Edwards described 'Varzob' as giving "a very pleasing two-tone effect, the outer segments buttercup yellow and lightly veined or feathered gray-green, contrasting with the deeper yellow, faintly veined inner ones; both are colored green towards the base."

At the heights of Sina in southeastern Uzbekistan *Crocus korolkowii* grew everywhere, but nowhere were there such large clumps as in Varzob. The plants in Sina are of the same pale color type as in Varzob, while those from Agalik and Aman-Kutan are a far better color, in my opinion.

Crocus korolkowii was also among E. A. Bowles's favorites and he selected several forms. Unfortunately, they were later lost. One of them I found unexpectedly in the collection of an Estonian gardener who had found it in a castle garden of a prewar landlord, and from there **'Dytiscus'** (Plate 188) was reintroduced in Britain. It is a very small-flowering form, producing up to 20 blooms per corm. The flowers are marked by deep purplish brown sepal backs and a very narrow golden yellow rim. E. A. Bowles (1952) wrote,

> [I] have selected, from collected roots, a very striking variety in which the back of outer segments is brown with a narrow margin of yellow. It reminded me so much of the handsome water beetle *Dytiscus marginalis* that I named it *Crocus korolkowii* var. *dytiscus*. It reproduces itself fairly truly from seed and seems a vigorous form.

'Dytiscus' is also a good increaser vegetatively. My Lithuanian friend Augis Dambrauskas had raised a stock to several hundred corms but just like me he lost all his corms in the winter of 2005–06. Since then I have received a few corms of 'Dytiscus' from my friends.

A photo of *Crocus korolkowii* **'Snow Leopard'** (Plate 191) was published in the September 1998 issue of the *Alpine Garden Society Bulletin*. Raised by John Grimshaw, this plant looked like *C. alatavicus* but with a purple throat. John's collection did not include *C. alatavicus* at the time the seeds were collected. In 2007 this plant received the Preliminary Commendation award from the AGS. The inner surface of the flower segments is ivory white with a small violet throat. The sepal outsides are of similar ivory white color, but the backs of the outer tepals are overlaid with a fine violet striation. Alan Edwards reports that he has never obtained seed from 'Snow Leopard'.

Czech gardeners gave me another similarly colored plant but with almost white sepal backs (only in the mid-zone are the backs slightly darker shaded) and with a whitish throat. Named *Crocus korolkowii* **'Albus'**, it was thought to be a hybrid between *C. korolkowii* and *C. michelsonii*.

A beautiful plant with white flowers and a deep purple throat once appeared suddenly among the seedlings of *Crocus korolkowii* in my garden. The sepal backs had a rather wide dark purplish stippled midzone (Plate 194). Most likely the pollen parent was *C. michelsonii*, because the seedling had a blue throat, though another possibility is the white-flowered *C. alatavicus*. I cannot be certain as I sowed seeds from an open-grown *C. korolkowii*, and the possible pollen parents were grown in the greenhouse and flowered much earlier. It might be better to regard this crocus as a white mutation of *C. korolkowii*. Unfortunately, it was lost during the horrible winter of 2005–06.

In April 2007 Sjaak de Groot from Holland photographed another white *Crocus korolkowii* specimen (Plate 195) among thousands of yellows in the Varzob gorge very far from the localities of *C. alatavicus* and *C. michelsonii*. This discovery confirmed that white forms sometimes appear in populations of this yellow-blooming species, but they are very rare.

Currently I am trying to stop selecting new plants from *Crocus korolkowii* seedlings, unless something exceptionally good shows up. It's hard to resist the temptation, however. Recently I marked a couple of seedlings with such wide and deep yellow flower segments that in overall looks they resembled some of the best *C. chrysanthus* forms. I still continue to sow its seeds in the hope of one day finding again a white-flowering specimen.

For many years I grew all my stocks of *Crocus korolkowii* only in the open garden where they flowered with the melting snow. In 1977, for example, the flowers showed up in early winter (December). I thought the plants were indestructible until the winter of 2002–03 when a December black frost destroyed the most vigorous crocuses; only a few corms survived from many varieties. The next blow came in the winter of 2005–06 when most of my outside-growing crocuses were wiped out. Now I plant several corms of all my crocus samples, regardless of the supposed hardiness, in the greenhouse in pots, though most of my *C. korolkowii* corms are still grown outside.

This crocus likes somewhat more clayey, rich, slightly acidic soil, but for pots I use my standard mix. Although *Crocus korolkowii* likes a dry and warm summer rest, sometimes I leave it in outside beds for two years before replanting. I haven't noticed any harm from such treatment, so it seems that this species is the most tolerant to the summer conditions of all Central Asian species.

The Strange Outliers

This group consists of species which Brian Mathew places into three small series, each of which included only one or two species. The plants have some very distinct features that separate them from the rest of the crocus species. With the exception of *Crocus fleischeri*, these species are not the easiest to grow in the garden. Most of them are grown in pots, although some have grown in my open garden for several years despite not being happy there.

1. Corm tunics with interwoven fibers .... *C. fleischeri*
1. Corm tunics fibrous and usually extended in the neck
    2. Leaves with wide white median stripe on surface, keel more or less rounded beneath
        3. Leaves curved on underside and with several shallow grooves but no keel, upper side with a wide silvery stripe .... *C. carpetanus*
        3. Leaves have two grooves but the keel is partly rounded with some extra shallow grooves on it .... *C. nevadensis*
    2. Leaves bright green with no median white stripe on the upper surface
        4. Flowers bright yellow .... *C. scardicus*
        4. Flowers deep purple, rarely white .... *C. pelistericus*

Crocus fleischeri J. Gay

PLATES 197 & 198

Crocus fleischeri ($2n = 20$) is named after Franz von Fleischer, who was the first to collect its corms. It is one of the most unusual crocus species with its bright yellow corms (only *C. boulosii* corms have a similar color) and a unique corm tunic which separates it from all the other crocuses. E. A. Bowles wrote that the plants "resemble those of *Iris (Gynandriris) sysirinchium* more than that of Crocus." The tunic is a wonderful piece of natural weaving, its fine fibers being so closely interwoven that they seem to be plaited.

Because of the characteristics of the corm tunic, Brian Mathew placed this species in the monotypic series *Intertexti*. According to the phylogenetic tree, it is a very close relative of *C. pestalozzae* from series *Biflori*. Although that species is morphologically similar to the other species in series *Biflori*, it is one of the most distinct in the series, growing on acid and alkaline soils and having the highest chromosome number (2*n* = 28).

In the wild *Crocus fleischeri* grows on open rocky hillsides or in sparse woods in southern and western Turkey as well as on a few East Aegean islands (Rhodes and Chios) from 750 to 1300 meters. I have mainly found it in forest clearings, where vegetation is removed to make way for electric lines, and in sparse spots between pine trees with less competing vegetation.

The shiny white flowers rarely have some shorter or longer deep purple midveins on the backs of the outer flower segments. The throat is variable—from yellow or orange to dark brown, sometimes even blackish. Usually the orange or brown base in the flower throat is surrounded by a wider or narrower diffused yellow zone. All types of base colors can be found within the same population. I mostly like the forms with darker throats.

The long many-branched bright orange stigma overtops the anthers and is very attractive. Although the flowers are comparatively small, they are very bright and really shine due to the color contrast between the flower segments and stigmatic branches. The traditional commercial form has somewhat larger corms, although they are still the smallest among the commercial crocuses; its flowers have a yellow throat.

In cultivation *Crocus fleischeri* increases very well by offsets, although the corms are small. The new corms remain firmly attached to the main corm and are enclosed in a tunic which they share with the main corm. Their separation requires extra work. In the wild the old tunics decay within a few years, freeing the replacement corms. I have found in the wild not more than one or two additional cormlets attached to the main corm. In cultivation, three to five cormlets are usual. Ole Sønderhausen reported specimens that formed stolons up to 10 cm long. At the bases of the stolons were young corms with thick fleshy roots. They somewhat resemble droppers in tulips (cited from B. Mathew).

Sønderhausen checked the pH of the soil at sites where *Crocus fleischeri* grows wild. It measured from 8.15 to 8.35. In other words, it was highly alkaline. I grow this crocus in the traditional soil mix and it looks quite happy here though the pH level is much lower (approximately 6.5).

Crocus fleischeri is a hardy species. I have been planting the traditional Dutch stock outside for many years and it suffered only slightly from the weather extremes of the recent winters. The forms from wild collections I still grow only in the greenhouse. *Crocus fleischeri* is very tolerant and seems to not need a dry, hot summer rest. I have left it in the outside garden without replanting for two or three seasons.

Crocus carpetanus Boissier & Reuter

PLATES 199 & 200

Crocus carpetanus (2*n* = 64) grows in the western end of the *Crocus* range. Unlike the typical crocus leaf which has a flattened keel between two grooves, the uniquely shaped leaves of this species have a rounded underside and resemble those of *Romulea* more than *Crocus*. Brian Mathew placed *C. carpetanus* and closely related *C.*

nevadensis in series *Carpetani*. In the phylogenetic tree both species stand side by side with romuleas, at some distance from other crocus species.

Crocus carpetanus is named after the Carpetani Montes in Central Cordillera of Spain where it was for the first time discovered. It occurs in central and northwestern Spain entering northern Portugal, where it can be found on granite formations in stony places and alpine meadows or sparse woods from 1200 to 2300 meters.

The flower color usually is pale lilac to whitish, often finely veined darker with a whitish or pale yellow throat. David Stephens (1999) comments: "It has almost crystalline quality to the colors which are quite variable from a good proportion of pure to near whites through tones of pink to nearly purples, with lots of bicoloreds."

My specimens are invariably pure white of quite strong texture and with a light yellow outside base color devoid of any veins. The throat is a very light, slightly grayish shade of creamy yellowish. The anthers are light yellow, and the stigma is divided into three white branches widely expanded and frilled at the tip and ending below the tips of anthers. Brian Mathew reports on a specimen with nearly white inner segments and outer ones entirely suffused on the outside with rosy purple. There can be dark veins at the base of outer segments as well. E. A. Bowles noted lilac-colored pistils in specimens with lilac-colored flower segments.

The leaves are very special, as mentioned above. Even a brief look at them allows immediate identification of the species. They have a wide white midvein on the channeled upper side, but the underside is rounded with a few shallow longitudinal grooves. The corm tunics are formed of pale, almost silvery, finely netted fibers.

In cultivation *Crocus carpetanus* has established a fairly bad reputation. E. A. Bowles reported that he lost all his plants both outside and in the bulb frame. I lost my samples grown outside within two to three years, but those planted in pots in the greenhouse grow well and are good increasers by splitting. In some seasons my plants set seed if an additional hand-pollination is provided but more often they produce empty seed capsules. My stock originates from Portugal and was collected near Port Torre at 1900 meters.

According to David Stevens (1999), bulbs in the wild often lie in the couple of centimeters of leaf litter under trees unlike the majority of crocus species. I would fear to pot them that shallow due to the winter frosts in my climate. I usually plant crocus corms approximately 5 cm deep and then cover them with 1.5 cm of stone chips.

Crocus carpetanus requires a non-alkaline soil and plenty of moisture in early spring, followed by a hot and dry summer. It has not been very difficult for me in cultivation, except when I forgot its soil pH needs one year and my plants developed light green leaves and the corms were much smaller than usual.

Crocus nevadensis Amo & Campo

PLATES 201 & 202

Crocus nevadensis ($2n = 28, 30$) is a close relative of *C. carpetanus* and in the phylogenetic tree the two species stand together. Named after the Sierra Nevada in southeastern Spain, *C. nevadensis* grows wild throughout eastern Spain and in North

Africa, where it is reported from northern Algeria and Morocco. It occurs on the Rif mountains and in the ridges of the Middle and High Atlas on mountain meadows and stony places from 500 to 2300 meters.

The flowers are white or very pale lilac, sometimes darkening with age and usually with violet veins on the exterior of the segments, which are often suffused pale green. The throat is whitish with a slight grayish shade and is pubescent (in *Crocus carpetanus* it is glabrous). The anthers are yellow and the style is shortly divided into three white branches ending below the tips of anthers. The leaves have the rounded shape of *C. carpetanus* but with two larger grooves on either side of the ridge. Although the leaves are somewhat halfway to a typical crocus leaf, they nevertheless are easily identifiable. The corm tunics are more or less parallelly fibrous extending into a brown neck.

Plants from North Africa are reported as easier in cultivation than those of Spanish origin. I have several stocks originating from Africa and from Spain, but I haven't noticed great differences in their behavior. My specimens from Trevenque in Andalusia, Spain, are paler in color. The sample from Morocco collected near Tizi-N-Ait Ovirra in the Middle Atlas, is brighter and the veins at the tips of the segments become bright lilac, but in all the other aspects the plants are quite similar in color.

For several years I grew *Crocus nevadensis* outside, but it did not do well nor did it flower. Then I grew it in the greenhouse where it turned out to be easy and now flowers abundantly every spring.

E. A. Bowles identified its greatest faults as the pale color and the narrow segments, describing it as an "interesting rather than a beautiful species." I cannot agree with this judgment. My samples have wide flower segments and I like the flower very much. Moreover, it seems to be easier in cultivation than *C. carpetanus*, requiring moist conditions in spring and a dry and warm summer rest.

Crocus scardicus and *C. pelistericus* form another couple of species which could be characterized even as "twins" if not the difference in the flower color. Both occur in Macedonia in the widest sense of the name—*C. scardicus* in the Republic of Macedonia (former Yugoslav section of the region of Macedonia) and entering Albania, and *C. pelistericus* on both sides of the border between the Greek province of Macedonia and the Republic of Macedonia. As is the case with the preceding couple of species, they both have very distinct leaves, differing from all other crocus species in the absence of the white stripe on the upper surface. Both are found in very specific growing conditions making them quite difficult in cultivation. Being very close, they hybridize quite easily. One of their offspring is named *Crocus* ×*gotoburgensis*.

Crocus scardicus Košanin

PLATE 203

Crocus scardicus (2*n* = 32, 34, 35, 36) is named after Mount Scardo to the west of Skopje, where the species flowers near melting snow in alpine grassland at 1700 to 2500 meters. It grows near streams on soils which never dry out in summer.

The flowers are very bright yellow or even orange with a purple base extending one-third up the length of the sepals. Rarely specimens occur with no purple color at the base of the tepals. In my specimens, the flower tube is entirely purple or purple striped with a slight extension of a purple zone on the flower segments. The combination of yellow and purple in the same flower is rather unusual among crocuses. The throat is generally white or yellow, sometimes light violet, but it is possible that the outside color shines through the tepal base. The anthers are yellow; the style is shortly divided yellow or orange usually at the same level or slightly surpassing the anthers. The corm tunics are finely fibrous-reticulated, forming a long persistent neck at the apex. Flowering-size corms are surprisingly small for a species that bears such large flowers.

This species is very little known in cultivation, and many attempts to introduce it failed before it was discovered that its natural habitat never dries out in summer. It flowers quite late in the season and is one of the last crocuses in my collection. When blooming ceases, a period of rapid leaf growth follows. The leaves remain green until late autumn. Usually the new shoot starts to elongate while the previous season's leaves are still green and active; before frost sets in the new shoots are only 2 to 3 cm below soil surface. Similarly, the roots of the previous season usually coexist with the new ones at the end of the season. Seeds ripen very late, too. In my collection, seed never ripens before the end of August or even September and is pushed up on a very long stalk.

I have not had much success growing *Crocus scardicus*, and my stock is not very diligent to flower. The greatest challenge is to maintain the right balance of moisture in the soil. One of the easiest ways to accomplish this is to keep the pots of *C. scardicus* in shallow trays filled with water all summer long. Then, from September till spring the pots must be kept drier and moved to the sand plunge. Perhaps a regular weekly or even more frequent watering and keeping the pots all the time in a plunge would be a better strategy. In the Scottish Rock Garden Club internet forum Dave Millward wrote, "Some growers stand the pot in a tray of water during the main growing season, but I have not found this necessary. I have also grown it in damper parts of the garden."

I replant *Crocus scardicus* in the second half of August and immediately put the corms in a fresh mix. Finding the best time to do this job isn't easy. I wait until the leaves start to die back. For this crocus I make a somewhat peatier substrate, adding a larger amount of peat moss to the usual soil mix. The result is a more acidic, moisture-retentive substrate that encourages active growth during the summer when all other crocuses have gone dormant. Dave Millward recommends John Innes compost with added 50 percent grit and some silica gel crystals to help retain moisture.

Crocus scardicus grows well outdoors at the Gothenburg Botanical Garden, where it abundantly flowers each spring. In the early years when I tried to grow it outside, I always lost it within several years.

Crocus pelistericus Pulevič

PLATES 204 & 205

Crocus pelistericus (2*n* = 34) was named in 1976 after Mount Pelister on the Yugoslavian side of Macedonia. It was described by V. Pulevič after many years of being misidentified as *C. veluchensis*. In some sites *C. veluchensis* can be found in great quantities with *C. pelistericus*, but the former always grows on drier soil around the periphery of *C. pelistericus* drifts and on south-facing slopes. On Mount Pelister it co-occurs with *C. scardicus* or at least both grow in the same mountain but in segregated populations with no reports about any hybrids between the two in the wild. It has also been found on the Greek side of the border on the Voras mountain range.

In everything but the flower color *Crocus pelistericus* is almost identical to *C. scardicus*. It occurs at similar elevations (over 2300 meters) and habitats (on marshy alpine meadows and fens). Both species flower in soil that is saturated, even partly flooded by late spring meltwater, and thus retains moisture through the summer months. Once it was even suggested that perhaps they are only color forms of one species, though both are well-separated geographically. In addition, *C. pelistericus* has wider and darker green leaves and a glabrous throat in the flowers, whereas the throat in *C. scardicus* is papillose.

Crocus pelistericus is famous for the unusual color intensity of its flowers, which are almost invariably deep rich violet. White flowers are very rare, at least in Greek populations. The flower tube is dark violet; the inner petals sometimes are somewhat lighter than the outer. The segments possess a reflective shine. This is one of the brightest purple crocuses. The attractive flower throat is white surrounded by lilac; in some specimens a very deep purple rayed zone borders the white center of the flower. The anthers are yellow; the style is shortly branched, white or light yellow to orange-yellow. Corms are very small and don't enlarge significantly under garden conditions but regardless of the size produce lavish foliage and huge flowers.

Crocus pelistericus and *C. scardicus* have similar growing requirements, though *C. pelistericus* has proved to be a more amenable and quite adaptable plant. I find it somewhat easier and more floriferous. For many years I grew it in the outside garden and never lost it completely. Now I grow it only in pots, giving it the same treatment I give *C. scardicus*. It readily sets seed even without additional hand-pollination. Flower production is rather sporadic; in some years it flowers sparsely, in others very abundantly, and sometimes it doesn't bloom at all.

Crocus ×*gotoburgensis* R. Rolfe

PLATES 206–208

Both *Crocus pelistericus* and *C. scardicus* have forms with same chromosome number. Being very close, they hybridize quite easily. Such hybrids were raised in the Gothenburg Botanical Garden by Henrik Zetterlund, who first crossed the two species in 1987. Later this hybrid was named *Crocus* ×*gotoburgensis*.

Henrik Zetterlund told me that there was no variation in the first-generation seedlings; all the flowers were colored like honey and blackberry. For this cross he used specimens of *C. pelistericus* from Karadjitsa Planina. When I first saw those hybrids, I named them "Rainbow," but subsequently they were renamed **'Ember'** by Robert Rolfe who obtained corms in 1994. In 2000 Rolfe wrote:

> [T]he general advice is to keep plants in growth for as long as possible, although this advice is not inviolable. . . . the corms . . . were out of their compost and rootless in October but did not suffer. They are grown in a frame with other more conventionally treated species and experience almost the same regime.

Possibly the corms did not suffer because of their hybrid vigor. I observed almost the same thing that Rolfe did in the several years that I grew *C.* ×*gotoburgensis* in the open garden. Only recently did I change my growing techniques and start to plant this hybrid in pots whereupon it became more floriferous.

Later this cross was repeated by other growers. The resulting first-generation seedlings were similar to those acquired in Gothenburg, but more yellowish with less violet in flowers.

In the second-generation seedlings the diversity is quite large. The best form seen by me has white flowers with a large deep purple base and a light violet upper third of the flower segments separated by a white midzone. Unfortunately this form increases very slowly, and a few years ago there were only three individuals. Some seedlings were almost pure white, others deep yellow with dark brown tepal tips. One other beauty was reddish lilac flushed over a yellowish base color, deepening at the tips to slightly brownish-tinged reddish-lilac tone—very difficult to describe. They really are among the best crocuses ever seen.

Yellow Fever

Gardeners usually attribute "yellow fever" to the daffodil blooming season, though it is equally applicable to crocuses which also have yellow flowers. Taxonomically these crocuses are very different, scattered among the numerous series according to Brian Mathew's classification. Several of them were described in previous chapters and won't be repeated here, but they are included in the key, which has a cross-reference to the appropriate chapter. Several species are quite similar if we only observe the flowers and therefore I was forced to include in this key one feature which I tried to avoid in all the previous—the presence or absence of the bracteole. I encountered problems while identifying some of the plants collected at sites far from the localities mentioned in floras, because some doubts remained about the occurrence of a certain species in the considered locality. Our knowledge about areas occupied by several species is still quite incomplete.

1. Corm tunics membranous, smooth with distinct basal rings .... see "The Fellowship of the Rings—Annulate Crocuses"
1. Corm tunics fibrous reticulated only at apex
   2. Style with three branches .... *C. herbertii*
   2. Style with 6 branches .... *C. olivieri* subsp. *istanbulensis*
1. Corm tunics finely or coarsely reticulated
   2. Corm tunics finely reticulated
      3. Leaves without prominent white midvein .... see "The Strange Outliers"
      3. Leaves with clearly visible white midvein on surface .... *C. cvijicii*
   2. Corm tunics coarsely reticulated
      4. Bract and bracteole present and visible
         5. Segments subacute, usually marked externally with purple-brown .... *C. angustifolius*
         5. Segments obtuse or rounded, unmarked on outside .... *C. ancyrensis*
         5. Segments obtuse or rounded, purplish yellow .... *C. ×paulineae*
      4. Bract only present, sheathing the tube .... *C. gargaricus*
1. Corm tunics parallelly fibrous
   6. Leaves 10–20 .... see "The Eastern Runners—Trio from Central Asia"
   6. Leaves up to 8
      7. Style with three branches
         8. Leaves up to 1.5 mm wide, corm tunics membranous without long neck at apex, splitting into coarse parallel fibers .... *C. sieheanus*
         8. Leaves 2.5–4 mm wide, corm tunics fibrous with long brown neck at apex .... *C. flavus* subsp. *flavus*
      7. Style with 6–10 branches
         9. Leaves one to four, corm with no long neck .... *C. olivieri* subsp. *olivieri*
         9. Leaves four to eight, corm with long brown neck
            10. Leaves 2.5–4 mm wide .... *C. flavus* subsp. *dissectus*
            10. Leaves 1–1.5 mm wide .... *C. flavus* subsp. *sarichinarensis*
      7. Style with more than 10 branches
         11. Leaves 3–6 mm wide, filaments glabrous............... *C. olivieri* subsp. *balansae*
         11. Leaves 0.5–3 mm wide, filaments papillose
            12. Leaves 5–8, gray-green, 0.5–1.5 mm wide, segments acute, 4–7 mm wide .... *C. graveolens*
            12. Leaves 2–4, shiny green, 1.5–3 mm wide, segments obtuse or rounded, 6–9 mm wide... *C. vitellinus*

Crocus flavus Weston

The most well-known species included in this chapter undoubtedly is *Crocus flavus*. It is the commonest yellow crocus grown in gardens, although in reality what is mostly grown is its hybrid with *C. angustifolius*, known as 'Golden Yellow' (its accepted name in the international checklist) and various other names such as 'Yellow Mammoth' and 'Grote Gele'. Although the parent species do not meet in the wild, they have been in cultivation since the sixteenth century and have often met in gardens.

'Golden Yellow' is sterile, and cytological studies confirmed its parentage. The hybrid is at least two hundred years old, so it isn't surprising that several mutations appeared in this time, resulting in different clones being offered under one or the other of its many names. It is an easy garden plant with large, somewhat pale yellow flowers of rounded form, in flower size almost equal to the large Dutch hybrids of *Crocus vernus*.

There is another sterile hybrid between these two species known in gardens under the name ***Crocus* ×*stellaris*** A. H. Haworth. It largely resembles *C. angustifolius*, having conspicuous stripes and feathering on the exterior of the yellow flowers, but is paler in the base color and the flowers are born higher. I grew it more than 25 years ago and then I noticed the very high increasing rate and hardiness of this hybrid. After the winter of 1978–79 this and *C. heuffelianus* were noted as the only crocuses blooming in my collection.

E. A. Bowles (1952) dedicates six pages to *Crocus flavus*, relating its history and describing the various forms known in cultivation at the time and I would like to only repeat the words of John Gerard cited by E. A. Bowles: "flowers of the most shining yellow color, seeming afar off to be a hot glowing cole [sic] of fire." He recommends planting the fertile wild form "in shrubberies where it may seed and spread freely under deciduous trees and shrubs," because the "rich orange color and early appearance of its flowers proclaim the typical form the best of its race."

Bright flowers attract the attention of birds. Brian Mathew reports an amazing recommendation in *Curtis' Botanical Magazine* (volume 1) regarding *C. flavus*: "We have succeeded in keeping these birds off, by placing near the object to be preserved, the skin of a cat properly stuffed." I fear that the use of such a remedy today would raise loud protests from various animal and wildlife protection organizations.

Numerous color forms of this species are described in the botanical literature as species, varieties, subvarieties, and so on. Usually they all refer to the garden forms of *Crocus flavus*. Brian Mathew in his monograph lists two subspecies of *C. flavus*. Just recently I found the third one in southern Turkey.

Crocus flavus subsp. *flavus*

PLATES 209 & 210

The type subspecies, *Crocus flavus* subsp. *flavus* ($2n = 8$), occurs in the Balkans and in Bulgaria and Romania in Europe and enters northwestern Turkey, where it grows in dryish grassland or open woodland from sea level up to 1000 meters. More to the south and east it is replaced by subsp. *dissectus*. It is generally a very vigorous plant that readily sets seed and in some places abundantly covers the ground with its grass-like leaves.

In the wild, as in cultivation, this species is fairly variable. The basic flower color ranges from pale lemon to deep orange-yellow depending on the population. In some populations the light-colored forms are usual, and elsewhere they are very rare or absent. More variable is the color of the flower tube, which ranges from nearly white or creamy yellow to brown striped and even violet stained. Sometimes the outer surface of segments is "clouded by brown markings," as noted by E. A. Bowles. Wherever I have come across this species in the wild I have found only bright yellow-colored specimens with a light flower tube.

Very special are the corm tunics which are membranous splitting at the base into parallel fibers and forming a long neck at the top. This "cap" persists for several years and its length depends on the depth at which the corm lies. It prepares the "road up" for the new shoot even in hard clayey soils if corms are left undisturbed for many years. It is not easy to take this long neck off at harvesting time, so strong is it. I usually use scissors. This neck makes identification of the species easier when a plant is found after flowering, although it can be misidentified with *Crocus antalyensis* in places where its area meets with subsp. *dissectus*.

In gardens the type subspecies is a vigorous plant, and in favorable conditions its leaves can grow as long as 60 cm. Usually it is hardy and can stay in the ground for several years without replanting, but after a very hard winter I lost all of my stocks as they were grown only outside. Some specimens increase rapidly by splitting but some never make any offsets. Subspecies *flavus* blooms abundantly and set seeds well.

Crocus flavus subsp. *dissectus* T. Baytop & B. Mathew

PLATES 211–213

Crocus flavus subsp. *dissectus* ($2n = 8$) replaces the type subspecies in the south and east directions. It resembles the type subspecies, except that its style is divided into six or more slender branches, its flowers are slightly smaller, and the plant is from higher elevations. I have come across it mostly in open pine or oak forests or in small clearings among shrubs, but it can be found in grass as well, between 500 and 1200 meters.

Once I found it on an open, quite rocky place and then I assumed that it was the blue-flowering *Crocus antalyensis*. I was not very pleasantly surprised the following spring when, after having taken such great pains to collect a few corms (and blisters on my palms) between stones and in brick-hard clay, this crocus turned out to be the bright yellow *C. flavus* subsp. *dissectus*. In the wild its corms lie quite deep (Plate 213). In one population which grew on very light sandy soil under large pine trees, the flowers emerged from approximately 25 cm depth.

This subspecies is less hardy than the type subspecies and I grow it only in the greenhouse. I came across the only albino of *Crocus flavus* in subsp. *dissectus* (before discovering subsp. *sarichinarensis*). Its flowers were of very nice ivory white shade, but as it was a very early and misty morning it was possible to photograph only its outside (Plate 212). It seems that subsp. *dissectus* produces offsets more slowly than subsp. *flavus* and the best way to increase the stocks is by seed.

Crocus flavus **subsp.** ***sarichinarensis*** Rukšāns, subsp. nov.

PLATES 214 & 215

> *Crocus flavus* subsp. *sarichinarensis* Rukšāns, subsp. nov., subsp. *dissectus* similes sed foliis angustatus (1–1.5 mm latus) et flores colores exterius pallidus suffusus bone differt.
>
> Typus: Turkey, Sarichinar-Dag, N slopes, 50 km W from Antalya, 1140 m. R2CV-035, deep clay at edge of meadow with deep leafy shrubs, rarely on small clearings between shrubs. Leg. J. Rukšāns, 16-03-2008. (GB, holo). Ic.: Crocuses: A Complete Guide to the Genus (Portland, OR, 2010), plates 214 and 215.

In spring 2008 I was traveling with friends through southwestern Turkey. Shortly before reaching Antalya city we stopped (altitude 1140 m) to take pictures of *Galanthus gracilis* and *Cyclamen trochopteranthum*. In dense shrubs I unexpectedly spotted a yellow crocus which in color looked just like some of the very common yellow cultivars of *Crocus chrysanthus* with brown speckled or striped sepal backs. It was getting quite late in the evening. Dark clouds forecasted heavy rains, and the conditions for taking a photo were poor; therefore I picked the flower for later examination in the hotel. Great was my surprise when I found the flower had a many-branched style rather than the trifid style of *C. chrysanthus*. The species had to be *C. flavus*, but I didn't know of any records putting it so far to the south in a coastal mountain ridge.

Fortunately, our schedule was not very tight and when the next morning greeted us with bright sunshine, it was not too difficult to persuade my colleagues to return to the site where the flower had been. We found that this crocus was very variable in color from almost pure white to quite deep yellow, but the outside of the outer segments was mostly covered with purplish, brownish, or grayish stripes, feathers, or stippling. White-flowered specimens had well-defined bright yellow throats. Plants grew on heavy and stony clay soils in clearings among deciduous shrubs but mostly in open spots near the border between meadows and shrubs. The corm tunics, the long neck, and the number of stigmatic divisions suggested *C. flavus* subsp. *dissectus*, but the generally paler color, very narrow leaves, and the location so far to the south from the earlier known southernmost locality of subsp. *dissectus* allowed me to presume that we had discovered a new subspecies of *C. flavus*.

I decided to name the new taxon *Crocus flavus* subsp. *sarichinarensis* after the mountain ridge in which it was found. Later I found that several specimens collected from this ridge during earlier trips and grown under the name *C. antalyensis*, at the first flowering turned out to be the new subspecies. Very similar plants were found by David Stephens east of Antalya, too (pers. comm.). The main difference between this subspecies and the other subspecies of *C. flavus* is its narrow leaves, only 1.0 to 1.5 mm wide. In the other two subspecies the leaves are 2.5 to 4 mm wide. There is a very little experience in growing the new subspecies, though by analogy with the neighboring *C. antalyensis* (in some places they form mixed populations) it most probably requires a dry summer rest and a deep planting.

Crocus olivieri J. Gay

Crocus olivieri (2*n* = 6), named after the French botanist G. A. Olivier who collected it for the first time on the Chios Island, is another very widespread and well-known bright yellow spring crocus species growing wild in the Balkans, southeastern Romania, southern Bulgaria, the Aegean Islands, and in northern, western, southern, and central Turkey. It is found in open or rocky grassy places or in woodland at 500 to 1400 meters. Its area mostly overlaps that of *C. flavus*, but it can also be found much further east. The two species are easy to tell apart by the stiffly, erect and more numerous leaves in *C. flavus* and the absence of the long brown "neck" of old cataphylls in *C. olivieri*.

The flower color of *Crocus olivieri* varies from pale lemon-yellow (though I have never seen such a shade) to bright yellow and even orange-yellow. Most of the flower tubes that I've seen are whitish or creamy, but occasionally I've seen grayish or brownish suffused tubes. Once I came across a specimen in which the brownish suffusion in the tube extended like a short "tongue" on the base of the flower segments.

Most of the plants encountered during my trips had few and wide leaves, but in vigorous specimens I observed up to five leaves. Their width varied greatly from 1.5 up to 7 mm, but mostly I saw plants with fewer though wider leaves. *Crocus candidus* is very close to *C. olivieri* but is easily separable by its white flowers. Three subspecies of *C. olivieri* are recognized by the shape of the corm tunics and the degree of style division

Crocus olivieri* subsp. *olivieri

PLATES 216 & 217

In the type, *Crocus olivieri* subsp. *olivieri*, the style is divided into six slender yellow or orange branches, slightly shorter or slightly exceeding the tips of anthers. The corm tunics are membranous, splitting at the base into parallel fibers. In general the flowers are yellow (Plate 216), but Ibrahim Sözen from Turkey found an incredibly beautiful specimen with pure white flowers (Plate 217), pale yellow anthers, and white stigmatic branches. The shape of flower segments clearly separates the white form of *C. olivieri* from *C. candidus*, a white-flowered relative of *C. olivieri*.

Subspecies *olivieri* is less hardy than could be presumed according to B. Mathew's monograph. I have lost stocks planted in the garden after every unfavorable winter, but in the greenhouse they survive. Perhaps the specimens from more northerly populations will be hardier. I have only samples gathered in Turkey.

Crocus olivieri* subsp. *balansae (J. Gay ex Maw) B. Mathew

PLATES 218 & 219

Crocus olivieri subsp. *balansae*, named after the French botanist B. Balansa who traveled in Turkey, is easily recognizable by its much-branched style which is divided into 12 to 15 branches. The exterior of the flower segments is usually striped or suffused purplish brown or mahogany. In the wild the subspecies occurs in western Turkey and on Samos and Chios Islands, growing on open hillsides and in scrub from sea level up to 1000 meters.

Although it is a plant of lower elevations, it seems to be more vigorous in gardens for this is the form mostly offered in bulb catalogs. I have never seen subsp. *balansae* in the wild and all my knowledge comes from forms in commerce, which didn't impress me much. But it is reported to be a very beautiful crocus, especially in its most bright forms. One of these is described by E. A. Bowles like this: "Its most striking form has a complete external suffusion of rich mahogany brown on the outer segments, so dark that brown madder shaded with purple-lake must be used to reproduce it in paint." Such color is seen in a cultivar named **'Chocolate Soldier'**; the glossy mahogany brown exterior of the tepals contrasts well with the bright orange yellow inside and fine threads of a deep orange stigma.

Like the type subspecies, subsp. *balansae* grows well planted in pots in a greenhouse.

Crocus olivieri* subsp. *istanbulensis B. Mathew

PLATE 220

The third subspecies, *Crocus olivieri* subsp. *istanbulensis*, was described by Brian Mathew in his monograph. It has a very limited area of distribution and is known only from the province of Istanbul, where it was found growing in dryish scrub and clearings at elevation from 150 to 170 meters. At present it is known only from a couple of localities highly endangered by the expanding Istanbul and there is a report about only some 300 specimens left in the wild.

Subspecies *istanbulensis* is distinguished from the other subspecies of *C. olivieri* by the corm tunics which become reticulated at the top. In subsp. *olivieri* the leaf arms are thick, consisting of two cell layers, and the keel is wide with a somewhat flared base. In subsp. *istanbulensis* the leaf arms are longer and thinner, consisting of one layer of cells, and the keel is narrow. In flower color it is somewhat variable from yellow to bright orange in the best forms. The style is six-branched.

This subspecies is very rare both in the wild and in cultivation, although from time to time it is offered in catalogs. I have bought it from two nurseries. One stock was heavily virus infected and regardless of repeated hand-pollination no seed capsule was formed. The other stock hasn't bloomed yet, and I cherish the hope that it will be true to name, at least its leaves look healthy. Like other forms of *Crocus olivieri*, this subspecies also flowers quite late.

Crocus ancyrensis (Herbert) Maw

PLATES 221–223

Crocus ancyrensis ($2n = 10$) is well known in gardens because it is offered by many nurseries and grown on a large scale in the Netherlands. It is very common in central Turkey and is named after the city of Ankara. It grows in oak scrub and open pine and fir forests, but most abundantly in open rocky places in mountain steppe vegetation from 800 to 1650 meters.

The flowers are generally bright yellow or orange, mostly with a whitish tube but in the best forms the tube is purplish and sometimes even the base of the outer segments is stained purple. The style is divided into three orange or reddish branches positioned around the tips of the yellow anthers. The filaments are short and glabrous. The corm tunics are strongly fibrous and coarsely reticulated.

This species is easy separable from *Crocus angustifolius* by the much rounder shape of the flower segments, which are devoid of purplish stripes. In cultivation a very floriferous clone is distributed under the name **'Golden Bunch'**; the flowers are indistinguishable from those of *C. chrysanthus* 'Uschak Orange', but start to flower one week after 'Uschak Orange'.

Some years ago I got a sample collected by a Czech traveler somewhat southward of the city of Tokat in the eastern part of *Crocus ancyrensis* area. The very beautiful flowers are brownish stippled throughout on the outside and a very deep yellow with brown shading on the inside (Plate 223). The throat is an even deeper yellow, slightly on the brownish orange side of the spectrum. In all other features it is a typical *C. ancyrensis*. It is a very good grower and increaser but I don't risk planting it outside. This form not only adds to the variability of *C. ancyrensis*, but it also is the only crocus known to me having such a color. Probably it is in color somewhat similar to var. *suffusus* mentioned by E. A. Bowles (1952). It never sets seeds with me.

In the wild *Crocus ancyrensis* increases freely by offsets as well as by seed and blooms right after the snow melt, but in Latvia it is a midseason bloomer both in the greenhouse and in the open garden. It is perfectly hardy, at least in its commercial form named 'Golden Bunch', which flowers particularly abundantly but doesn't set seed. *Crocus ancyrensis* is a good grower in pots, but doesn't need a very hot summer rest. It has managed in the garden for several years without replanting.

Crocus ×*paulineae* Pasche & Kerndorff

PLATES 265 & 306

Near Lake Abant in Bolu, Turkey, *Crocus ancyrensis* grows in mixed populations with the blue-flowering *C. abantensis*. The blooming of *C. ancyrensis* starts somewhat earlier, but it corresponds with that of *C. abantensis* and sometimes they intercross. The resulting hybrid was named by Erich Pasche and Helmut Kerndorff *C.* ×*paulineae* in honor of Pauline M. Dean, who has made drawings of all the new *Crocus* taxa discovered by them since 1991.

The flowers of this hybrid between a yellow and a blue species are somewhat variable, usually with a yellow base color, faintly suffused with purple. Sometimes the outer segments are darker than the inner. Both parent species belong to series *Reticulati* so their mutual hybridization isn't a big surprise, but it is not a very common case either.

I have been to the heights above Lake Abant four times and have found only one specimen of a clearly hybrid origin. Nice bicolored yellow-purple seedlings appeared among my open-pollinated seedlings of *C. ancyrensis* grown in greenhouse side by side with *C. abantensis*.

Plate 116. *Crocus chrysanthus* (JATU-068) on Gembos yaila in Turkey.

Plate 118. *Crocus chrysanthus* with purple flower tube from Gencek in Turkey.

Plate 117. *Crocus chrysanthus* on Gembos yaila in Turkey.

Plate 119. *Crocus chrysanthus* 'Macedonian Ivory'.

Plate 120. *Crocus chrysanthus* 'Sunspot'.

Plate 121. *Crocus chrysanthus* (R2CV-053) with black stigma on Gembos yaila in Turkey.

Plate 122. *Crocus chrysanthus* (R2CV-054) with black connective of anthers on Gembos yaila in Turkey.

Plate 123. *Crocus chrysanthus* 'Gundogmus Bronze'.

Plate 125. *Crocus chrysanthus* 'Goldmine', a double-flowered cross.

Plate 124. *Crocus* 'Snow Crystal', a hybrid between *C. biflorus* and the *C. chrysanthus*, was selected by the author.

Plate 126. *Crocus* 'Jūrpils', a seedling from *C. biflorus* and *C. chrysanthus* hybrid group, was raised by Juris Egle, Latvia.

Plate 127. *Crocus almehensis* in the wild at Almeh in Iran, from side. Photo: John Ingham.

Plate 128. *Crocus almehensis* in the author's collection.

Plate 129. Yellow form of *Crocus danfordiae* (KPPZ-304).

Plate 130. Blue form of *Crocus danfordiae* (R2CV-047) near Beyşehir, Turkey

Plate 131. Blue form of *Crocus pestalozzae*.

Plate 132. *Crocus cyprius*. Photo: M. Kammerlander.

Plate 133. *Crocus cyprius*.

Plate 134. *Crocus leichtlinii* (KPPZ-144) from between Siverek and Diyarbakir, Turkey. Photo: M. Kammerlander.

Plate 135. *Crocus leichtlinii.* Photo: John Lonsdale.

Plate 136. Another form of *Crocus leichtlinii* (KPPZ-163), from Hop Gecidi near Mardin, Turkey. Photo: M. Kammerlander.

Plate 137. *Crocus kerndorffiorum* (HKEP-9010).

Plate 138. *Crocus paschei* (HKEP-9034) from near Karamanmaras, Turkey. Photo: M. Kammerlander.

Plate 139. *Crocus aerius* from Zigana Pass in northeastern Turkey.

Plate 140. *Crocus adanensis.*

Plate 141. *Crocus tauricus* on Tschatir-Dag yaila in Crimea, Ukraine.

Plate 142. White form of *Crocus tauricus* from Tschatir-Dag yaila in Crimea, Ukraine.

Plate 143. Reddish form of *Crocus tauricus.*

Plate 144. *Crocus biflorus* var. *parkinsonii*, an old garden cultivar.

Plate 145. *Crocus biflorus* subsp. *biflorus*. Photo: John Lonsdale.

Plate 146. White form of *Crocus biflorus* subsp. *biflorus*. Photo: Ibrahim Sözen.

Plate 147. *Crocus biflorus* subsp. *alexandri*, closed flowers. Photo: M. Kammerlander.

Plate 148. *Crocus biflorus* subsp. *alexandri*, open flowers Photo: M. Kammerlander.

Plate 149. White form of *Crocus biflorus* subsp. *alexandri*. Photo: Ibrahim Sözen.

Plate 150. *Crocus biflorus* subsp. *weldenii* from Slovenia.

Plate 151. *Crocus biflorus* subsp. *weldenii* 'Albus'.

Plate 152. *Crocus biflorus* subsp. *stridii* with blackish maroon anthers from Cortiatis village in Thessalonika, Greece. Photo: John Lonsdale.

Plate 153. *Crocus biflorus* subsp. *stridii* with yellow anthers from Cortiatis village in Thessalonika, Greece. Photo: John Lonsdale.

Plate 154. *Crocus biflorus* subsp. *adamii* at Zangezur Mountains in Armenia. Photo: Sjaak de Groot.

Plate 155. *Crocus biflorus* subsp. *adamii* near Vanadzor in Armenia. Photo: Zhirair Basmajyan.

Plate 156. *Crocus biflorus* subsp. *adamii* from Vanadzor in Armenia.

Plate 157. *Crocus biflorus* subsp. *artvinensis* (HKEP 9359) from Artvin, Turkey. Photo: M. Kammerlander.

Plate 158. *Crocus biflorus* subsp. *artvinensis* (HKEP-9359).

Plate 159. *Crocus biflorus* subsp. *isauricus* (JATU-073) at Aldurbe yaila near Akseki in Turkey.

Plate 160. *Crocus* sp. collected as *C. biflorus* subsp. *isauricus* (RUDA-008) near Ermenek in Turkey. Possibly a new subspecies from the *C. biflorus* group. Some plants are without black connective of anthers.

Plate 161. *Crocus biflorus* subsp. *isauricus* at Gembos yaila grows side by side with *C. chrysanthus* and *Eranthis hyemalis.*

Plate 162. A very unusually colored natural hybrid between *Crocus biflorus* subsp. *isauricus* and *C. chrysanthus* at Gembos yaila.

Plate 163. White form of *Crocus biflorus* subsp. *punctatus.*

Plate 164. Blue form of *Crocus biflorus* subsp. *punctatus* (NJ 90-5A).

Plate 165. *Crocus biflorus* subsp. *crewei.* Photo: Erich Pasche.

Plate 166. *Crocus biflorus* subsp. *nubigena* (JP86-09).

Plate 167. *Crocus biflorus* subsp. *nubigena.* Photo: John Lonsdale.

Plate 168. *Crocus biflorus* subsp. *pseudonubigena* (KPPZ-108).

Plate 169. *Crocus biflorus* subsp. *pseudonubigena* (KPPZ-131). Photo: M. Kammerlander.

Plate 170. *Crocus biflorus* subsp. *caricus* (R2CV-019). Photo: Hendrik van Bogaert.

Plate 171. Possibly a new subspecies of *Crocus biflorus* (BATM-385) by some features resembling subsp. *caricus.*

Plate 172. *Crocus biflorus* subsp. *pulchricolor* and *C. herbertii* at Ulu-Dag in Turkey.

Plate 173. Best form of *Crocus biflorus* subsp. *pulchricolor* (MP 81-02) from the author's collection.

Plate 174. *Crocus biflorus* subsp. *tauri* (LST-185) from Sakaltutan mountain pass.

Plate 175. *Crocus biflorus* subsp. *tauri* in garden.

Plate 176. *Crocus* species close to *C. biflorus* subsp. *tauri* (WHIR-100) from Iran.

Plate 177. *Crocus biflorus* subsp. *atrospermus* (JATU-030).

Plate 178. *Crocus biflorus* subsp. *leucostylosus* (HKEP-0214). Photo: M. Kammerlander.

Plate 179. *Crocus biflorus* subsp. *caelestis* (HKEP-0409). Photo: M. Kammerlander.

Plate 180. *Crocus biflorus* subsp. *fibroannulatus* (HKEP-9361).

Plate 181. *Crocus biflorus* subsp. *munzurense* (HKEP-9911). Photo: M. Kammerlander.

Plate 182. *Crocus michelsonii* from Turkmenistan.

Plate 184. *Crocus michelsonii* 'Odissey' raised by Leonid Bondarenko, Lithuania.

Plate 183. *Crocus michelsonii* from Ala-Dag in Iran. Photo: M. Kammerlander.

Plate 185. *Crocus alatavicus* above Great Almaty Lake in Kazakhstan. Photo: Sjaak de Groot.

Plate 186. *Crocus alatavicus* from Bishkek, Kyrgyzstan. Photo: M. Kammerlander.

Plate 187. *Crocus korolkowii* 'Dark Throat'.

Plate 188. *Crocus korolkowii* 'Dytiscus' was raised by E. A. Bowles and is still in cultivation.

Plate 189. *Crocus korolkowii* 'Golden Nugget'.

Plate 190. *Crocus korolkowii* 'Lucky Number'.

Plate 191. *Crocus korolkowii* 'Snow Leopard'.

Plate 192. *Crocus korolkowii* 'Varzob'.

Plate 193. *Crocus korolkowii* 'Yellow Tiger' is now grown in large numbers in Jan Pennings's nursery in the Netherlands.

Plate 194. A white seedling of *Crocus korolkowii* that appeared in the author's collection.

Plate 195. A rare white-flowered *Crocus korolkowii* in the wild in the Varzob valley of the Hissar Mountains in Tajikistan. Photo: Sjaak de Groot.

Plate 196. Very unusual apricot-toned *Crocus korolkowii* specimen from the Varzob valley in Tajikistan. Photo: Sjaak de Groot.

Plate 197. *Crocus fleischeri* (JATU-018).

Plate 198. *Crocus fleischeri* from between Kale and Muğla in Turkey.

Plate 199. *Crocus carpetanus.*

Plate 200. Variability of *Crocus carpetanus* in the wild. Photo: Rafa Diez Domingues.

Plate 201. *Crocus nevadensis* in the author's collection.

Plate 202. Darker form of *Crocus nevadensis*.

Plate 203. *Crocus scardicus*.

Plate 204. *Crocus pelistericus*, from side.

Plate 206. *Crocus ×gotoburgensis*.

Plate 205. *Crocus pelistericus*, from top.

Plate 207. Second-generation seedling of *Crocus ×gotoburgensis*.

Plate 208. Second-generation seedling of *Crocus ×gotoburgensis* 'Henrik'.

Plate 209. *Crocus flavus* subsp. *flavus*.

Plate 210. *Crocus flavus* subsp. *flavus* near the road to Ulu-Dag, Turkey.

Plate 211. *Crocus flavus* subsp. *dissectus* (LST-064).

Plate 212. Creamy white specimen of *Crocus flavus* subsp. *dissectus* (R2CV-064).

Plate 213. *Crocus flavus* corms lie very deep in the soil (LST-069)

Plate 214. One color form of *Crocus flavus* subsp. *sarichinarensis* (R2CV-035).

Plate 215. Another color form of *Crocus flavus* subsp. *sarichinarensis* (R2CV-036).

Plate 216. *Crocus olivieri* subsp. *olivieri* (JATU-002) from near Lake Abant, Turkey.

Plate 217. White form of *Crocus olivieri* subsp. *olivieri* found in the wild by Ibrahim Sözen. Photo: Ibrahim Sözen.

Plate 218. *Crocus olivieri* subsp. *balansae*. Photo: M. Kammerlander.

Plate 220. *Crocus olivieri* subsp. *istanbulensis*.

Plate 219. *Crocus olivieri* subsp. *balansae*.

Plate 221. *Crocus ancyrensis* at the pass above Lake Abant, Turkey.

Plate 222. *Crocus ancyrensis* (LST-114) in a garden.

Plate 223. Brownish form of *Crocus ancyrensis*.

Plate 224. *Crocus angustifolius* in the author's collection.

Plate 225. Selection from seedlings of *Crocus angustifolius*—'Berlin Gold'.

Plate 226. Bronze-toned hybrid of *Crocus angustifolius.*

Plate 227. *Crocus ×leonidii* 'Alionka'.

Plate 228. *Crocus ×leonidii* 'Ego'.

Plate 229. *Crocus ×leonidii* 'Janis Ruksans'.

Plate 230. *Crocus ×leonidii* 'Little Amber'.

Plate 231. A very unusually colored form of *Crocus cvijicii*. May be even hybrid.

Plate 232. *Crocus cvijicii.*

Plate 233. *Crocus cvijicii* 'Cream of Creams'.

Plate 234. White seedlings of *Crocus cvijicii.*

Plate 235. *Crocus cvijicii* hybrid with *C. veluchensis.*

Plate 236. *Crocus gargaricus* is one of the brightest orange crocuses.

Plate 237. By color *Crocus herbertii* can compete only with *C. gargaricus*.

Plate 238. *Crocus herbertii* forms short white stolons at the end of which develop small cormlets.

Plate 239. *Crocus graveolens* in the wild near Pozanti in Turkey. Photo: David Millward.

Plate 240. *Crocus graveolens*. Photo: John Lonsdale.

Plate 241. One color form of *Crocus graveolens* (RIGA-115).

Plate 242. Another color form of *Crocus graveolens* (RUDA-104).

Plate 243. *Crocus vitellinus* from southern Lebanon. Photo: Oron Peri.

Plate 244. *Crocus vitellinus* in the wild near Gülnar in Turkey (JJVV-034).

Plate 245. Cultivated form of *Crocus vitellinus*.

Plate 246. *Crocus sieberi* from Omalos plain in Crete, Greece.

Plate 247. Color form of *Crocus sieberi* selected by Franz Hadacek from Austria. Photo: Franz Hadacek.

Plate 248. *Crocus sieberi* 'Cretan Snow'.

Plate 249. *Crocus* 'George', a mutation of *C.* 'Hubert Edelsten', was selected and named by Willem van Eeden in honor of George Rodionenko.

Plate 250. *Crocus atticus* subsp. *atticus*.

Plate 251. *Crocus atticus* subsp. *atticus* 'Bowles' White'.

Plate 252. *Crocus atticus* subsp. *sublimis* from Mount Parnassus in Greece.

Plate 253. *Crocus atticus* subsp. *sublimis* 'Tricolor'.

Plate 254. *Crocus atticus* subsp. *sublimis* 'Michael Hoogs Memory'.

Plate 255. *Crocus atticus* subsp. *nivalis* (JP-8802). Photo: M. Kammerlander.

Plate 256. *Crocus rujanensis*.

Plate 257. *Crocus dalmaticus*.

Plate 258. *Crocus veluchensis.*

Plate 259. White forms of *Crocus veluchensis.*

Plate 260. *Crocus reticulatus* from Moldavia (formerly Bessarabia).

Plate 261. *Crocus reticulatus* from Gülek Pass in Turkey.

Plate 262. *Crocus hittiticus* (SASA-022).

Plate 263. *Crocus abantensis* at Lake Abant in Turkey.

Plate 264. *Crocus abantensis* in garden.

Plate 265. *Crocus abantensis* 'Azkaban's Escapee'.

Plate 266. *Crocus* ×*paulineae* at Lake Abant.

Plate 267. *Crocus etruscus* (CG-8315).

Plate 268. *Crocus baytopiorum*.

Plate 269. *Crocus vernus* from Croatia.

Plate 271. *Crocus heuffelianus* 'Carpathian Wonder'.

Plate 270. *Crocus heuffelianus* in Slovakia. Photo: Vladimir Ježovič.

Plate 272. Hybrid between *Crocus heuffelianus* and *C.* ×*cultorum* 'National Park'.

Plate 273. *Crocus heuffelianus* subsp. *scepusiensis*.

Plate 274. *Crocus albiflorus* (TCH-3608) from Slovenia.

Plate 275. *Crocus tommasinianus* 'Bobbo'.

Plate 276. *Crocus tommasinianus* 'Lilac Striped'.

Plate 277. *Crocus tommasinianus* 'Pictus'.

Plate 278. *Crocus tommasinianus* 'Ruby Giant'

Plate 279. *Crocus tommasinianus* seedling in John Grimshaw's garden.

Plate 280. Pinkish-toned *Crocus tommasinianus* seedling selected by John Grimshaw.

Plate 281. Hybrid between *Crocus tommasinianus* and *C.* ×*cultorum* 'Yalta'.

Plate 282. *Crocus kosaninii*.

Plate 283. *Crocus malyi* (CEH-519).

Plate 284. *Crocus malyi* hybrid with unknown species raised in the Gothenburg Botanical Garden, Sweden.

Plate 285. One color form of *Crocus versicolor* selected in the author's nursery.

Plate 286. Another color form of *Crocus versicolor* selected in the author's nursery.

Plate 287. *Crocus imperati* from Amalfi, Italy.

Plate 288. *Crocus imperati* var. *reidii.*

Plate 289. *Crocus suaveolens.*

Plate 290. *Crocus corsicus.*

Plate 291. Form of *Crocus minimus* grown by large commercial nurseries in the Netherlands.

Plate 293. *Crocus minimus* 'Bavella'.

Plate 294. *Crocus antalyensis* at Ekizce yaila near Antalya in Turkey (JJVV-051).

Plate 292. *Crocus minimus* from Mount Limbara in Sardinia. Photo: M. Kammerlander.

Plate 295. Dark form of *Crocus antalyensis* (JJVV-052).

Plate 296. *Crocus antalyensis* subsp. *striatus* (RUDA-006).

Plate 297. *Crocus candidus* (R2CV-007).

Plate 298. *Crocus candidus*. Photo: John Lonsdale.

Plate 299. *Crocus candidus* yellow mutation found wild in Turkey by Ibrahim Sözen. Photo: Ibrahim Sözen.

Plate 300. *Crocus candidus* 'Little Tiger', a hybrid between *C. candidus* and *C. olivieri*.

Plate 301. *Crocus* sp. found where *C. kotschyanus* and *C. cancellatus* areas merge. Photo: Kees Jan van Zwienen.

Plate 302. May be a new crocus subspecies from *C. biflorus* group in southern Turkey, east of Antalya. Photo: Dave Millward.

Plate 303. New, still unnamed subspecies of *Crocus biflorus* close to subsp. *crewei* (JJVV-038).

Plate 304. *Crocus biflorus* subspecies from Iran (WHIR-163).

Plate 305. *Crocus biflorus* subspecies close to subsp. *tauri* from Malatya in Turkey (JRRK-075).

Plate 306. *Crocus* ×*paulineae*, raised in the author's nursery.

Plate 307. Natural hybrids between *Crocus chrysanthus* and *C. biflorus* subsp. *isauricus* at Gembos yaila near Akseki in Turkey.

Crocus angustifolius Weston

PLATES 224–226

Crocus angustifolius ($2n = 12$) has a corm tunic similar to that of *C. ancyrensis* and grows wild on the opposite (northern) coast of the Black Sea. It is better known in gardens under its synonym name *C. susianus* or as "Cloth of Gold," and has been cultivated since the sixteenth century. In Russian botanical literature it is still mostly named as *C. susianus*. In the wild it is distributed only in southern Ukraine and Crimea where it grows on open hillsides or in open woods from 200 to 1500 meters.

I came across *Crocus angustifolius* during one of my first trips to Crimea, where it abundantly flowered at the end of March along serpentine roads winding down to the Black Sea near Oreanda. It flowered in narrow strips bordering the asphalt pavement; the strips were almost devoid of other plants. The population varied in the amount of purplish brown striping on the back of the sepals; in some specimens it was almost absent. Compared to the tradition commercial form, this form's greatest advantage is its fertility. The form grown by me under the name 'Cloth of Gold' never sets seed but readily increased by offsets. The maximum number noted by me came to 10 new corms for one planted corm, but the average was 4.9. The wild forms are much lazier in this aspect.

Although *Crocus angustifolius* is similar to *C. ancyrensis*, both species are easy to separate for the flowers of *C. angustifolius* are starrier in appearance, with pointed tepals, and in full sun they open widely like small yellow stars adpressed to the ground. The flower tube usually is striped or entirely purple and, as a rule, the sepal outsides are marked with purplish brown stripes, a feature never observed in *C. ancyrensis*. In some wild forms the space between the stripes is filled with a brown tinge, although there are forms in which the brown stripes are absent.

A pure yellow form was selected by the German gardener Werner Wolf who named it 'Gold' but later renamed it **'Berlin Gold'**. Its flowers are of brightest yellow when widely opened, but I mostly like it with closed buds, which are somewhat paler in shade, pastel toned with lighter backs on the outer segments. The flower tube in this selection is greenish white becoming creamy shortly below where the flower segments split. The shape of the flower tepals seems to be somewhat more rounded and I fear that it probably is a hybrid with *C. ancyrensis*, because it was selected from garden-raised seedlings, not from wild material, and it never sets seed.

A hybrid between *Crocus angustifolius* and an unknown species is offered by some companies as *C. angustifolius* **'Bronze Form'**. In flower shape 'Bronze Form' closely resembles *C. angustifolius*, but the base color is somewhat bronze-shaded with a diffused lilac feathering on the outside of the segments (Plate 226). Its corm tunics are fibrous, but the fibers are mostly parallel with only occasional junctions and a few meshes. It is impossible to judge what the other parent of this hybrid is. It is sterile and never sets seed.

Crocus angustifolius is a good grower in the open garden but requires a warm summer period. Therefore it is essential to plant it in a well-drained sunny spot. It is a good grower in pots as well. I use a traditional mix.

Crocus ×leonidii Rukšāns

PLATES 227–230

Crocus ×leonidii Rukšāns, hybridus hortus inter *Crocus reticulatus* Steven ex Adam atque *Crocus angustifolius* Weston quasi intermedia et ex hybridatione harum specierum orta in Horto Jānis Rukšāns. Flores colores *Crocus angustifolius* approximatus sed habitu similis *Crocus reticulatus.*

Typus: cultivated material from Jānis Rukšāns garden (GB, holo, ex culturae in Horto Jānis Rukšāns, 16.03.2009). Ic.: Crocuses: A Complete Guide to the Genus (Portland, OR 2010), plates 227–231.

In the garden I always planted *Crocus angustifolius* alongside its bluish neighbor *C. reticulatus* (though I have never seen them growing together in the wild). Every year I collected seeds only of *C. reticulatus* and was very surprised when among its seedlings suddenly appeared plants with yellow flowers. In color they resembled *C. angustifolius* very much, but in overall appearance their shape was more of that of *C. reticulatus.*

My Lithuanian friend Leonid Bondarenko selected and named several clones from these seedlings. One of them was named in my honor **'Janis Ruksans'** (Plate 229). Its flowers are golden-yellow throughout with wide, almost convergent reddish brown stripes on the sepal backs. It blooms very early and abundantly, producing 8 to 10 flowers from a corm, and the blooming continues for a long time. This plant was registered with the KAVB in 2004. (The KAVB, or Royal General Bulb Growers Association, acts as the International Cultivar Registration Authority for Crocus and other bulbs.)

The earliest-blooming seedling, named **'Early Gold'**, can form up to 13 flowers from one corm and has a surprisingly high rate of increase. The lightest colored form from this hybrid series, with creamy yellow basic color and very prominent reddish brown stripes on the back of the sepals is named **'Nida'**. It, too, has a very high rate of increase. **'Ego'** has the largest flowers (Plate 228).

Several of Leonid's selections are now grown by Dutch nurseryman Jan Pennings and soon they will be available in the trade. I think that those hybrids are worth a specific name and I want to propose the name *Crocus ×leonidii* to honor Leonid's work in selecting, multiplying, and introducing these beautiful and vigorous crocuses.

Crocus cvijicii Košanin

PLATES 231–235

Much rarer in gardens than *Crocus angustifolius*, though even hardier, is *C. cvijicii* ($2n$ = 18, 19, 20) from southern Yugoslavia, northern Greece, and Albania. About this plant's name E. A. Bowles wrote: "I have never discovered how this name should be pronounced, whether it is better to imitate a sneeze or, as a witty friend of mine put it, 'to play it on the violin'." The species honors Yugoslavian geologist Jovan Cvijič (pronounced "shvee-yeech"). *Crocus cvijicii* grows in high mountain meadows on limestone formations where it flowers near melting snow from 1800 to 2500 meters, sometimes in clearings of subalpine forests.

The flowers of the most widely cultivated forms of the species are of the brightest yellow, but in the wild they can vary from rather pale to very bright and rarely even white. According to Brian Mathew, the paler yellow forms come from Yugoslavian and Albanian populations, while those from Greece are deeper yellow toned. The flower tube usually is short, whitish or yellow but in some forms can be lilac. I have observed only lightly colored flower tubes in my plants and a few with light purplish stripes. Because of the short tube, the flowers usually open at soil level and only later are pushed higher. The closest relative, *Crocus veluchensis*, is easily separable not only by the purple color, but also by the longer flower tube. The style divides, more or less at the level of the anther tips, in three widely expanded branches, the tips of which are yellow to bright orange. Only in pure albinos can the style be white.

The corm tunics are formed of thin fibers and are finely reticulated. In the wild even the flowering-size corms are very small at only 5 to 10 mm in diameter, but their size significantly increases in cultivation. I once ordered two dozen corms from a Czech gardener and was very disappointed with their small size. When I complained, I was told that the corms had been collected in the wild (I couldn't imagine that when I ordered them as I never sell plant material from the wild) and were of flowering size. And he was right. They all bloomed perfectly in spring.

Most of my *Crocus cvijicii* samples are lighter or deeper bright yellow. The form which I got under the name var. *alba* turned out to have very beautiful light creamy yellow flowers. It was so nice that I named it **'Cream of Creams'** (Plate 233). *Crocus cvijicii* readily sets seed which has a good rate of germination if sown immediately after harvesting. Among the seedlings of my creamy yellow form unexpectedly appeared a beautiful plant with a lilac flush over the slightly creamy white background color, clearly showing its hybrid origin with *C. veluchensis* as the pollen parent, which usually blooms at the same time (Plate 235).

Alan Edwards has taken a picture of a purest white specimen (see *AGS Bulletin* 66, no. 3). He wrote that the whole morning they had walked on the slopes of Mount Vermion "among millions of *C. cvijicii* expecting to find some variation in color, but surprisingly there was none, save that we were lucky enough to find just three albinos with purple stained tubes." Among my seedlings of 'Cream of Creams' appeared several purest white specimens with a white flower tube, white throat, and white filaments; only the anthers and stigma are yellow (Plate 234).

Crocus cvijicii is completely hardy and has endured even the hardiest winters in the open garden. I grow it in pots as well, mostly to protect it from rodents which every winter shorten the list of my collection by several stocks. Because the species comes from high elevations, it doesn't need hot and dry summer conditions. In fact, it can suffer if the soil in the pot during the summer dries out completely. I have found it better in my climate to sow seeds of *C. cvijicii* in open beds than in pots, probably because of a better moisture regime. The seeds germinate well and first flowers appear after three to four years.

In the garden *Crocus cvijicii* is one of the latest crocuses, and in the greenhouse only a few species (*C. minimus*, *C. pelistericus*, and *C. scardicus*) bloom later. As its flowers emerge well before leaves, I'm always a little nervous—has my favorite survived the winter? 'Cream of Creams' blooms even later than the yellow forms.

It is quite surprising that such a good plant is still so rare. Sometimes under this name are sold other yellow crocuses, mostly *Crocus chrysanthus*.

Crocus gargaricus Herbert

PLATE 236

The status of two other golden yellow crocus species has changed several times. These two are *Crocus gargaricus* and *C. herbertii*. Known among gardeners as non-stoloniferous and stoloniferous forms or subspecies of *C. gargaricus*, they were joined by Brian Mathew in his monograph but later again separated by him. Subsequent phylogenetic research has confirmed that both are different not only at the subspecies level but are two easily separable species. I don't know how to differentiate them by flowers; possibly those of *C. gargaricus* are somewhat larger when grown side by side, but all doubts disappear as soon as you see their corms. In the phylogenetic tree, *Crocus gargaricus* is placed in a well-supported clade among *C. cancellatus*, species of series *Speciosi*, another reticulated yellow-colored species, and several taxa from series *Biflori*, confirming its distance from *C. herbertii*.

Crocus gargaricus ($2n = 30$) was described from Mount Gargarus of the Kaz-Dag mountains in western Turkey where it grows on damp pastures and sparse pine woodlands from 1300 to 2300 meters, flowering just after snow melt. The moisture from melting snow at higher elevations keeps the soil wet for many weeks. The species is also found in mountains near Gōktepe in the province of Muğla. On the Kaz-Dag the environment is significantly more moist as the conditions there are more typically alpine.

This small and dwarf species has such a bright yellow, even orange flower that it belongs to the best early spring-blooming crocuses. Perhaps it belongs to species with the most orange-toned flowers, in shade surpassing all the other crocuses known to me. The style is three-branched, usually orange but sometimes yellow, dividing approximately at the same level as the tips of anthers. The throat is even darker orange. It seems that there is a very little variation in flower color although a lemon yellow form has been recorded and even named by William Herbert as var. *citrinus*. I had a few plants with a yellow throat (of the same color as the upper part of the segments). The corms are comparatively small but larger than in *C. herbertii* and covered by coarsely reticulated tunics.

Crocus gargaricus is a perfectly hardy plant, growing well in my garden. As a high mountain plant from moist meadows it doesn't need summer baking. I move all my pot-grown samples outside when the night frosts are not so hard anymore. The species readily sets seed and increases by splitting. It seems somewhat more tolerant of a little drier summer conditions than *C. herbertii*. It is easy in the open as well and sometimes even self-seeds here.

Crocus herbertii (B. Mathew) B. Mathew

PLATES 172, 237 & 238

Crocus herbertii ($2n = 30$) may have somewhat smaller flowers than *C. gargaricus*, but they are of the same brightest orange-yellow color (Plate 237). The species is distributed in the Ulu-Dag mountains in northwestern Turkey and is very similar to *C. gargaricus* except that its corms form short white stolons with very small grainlike cormlets at their ends (Plate 238). The specific name honors the great botanist William Herbert who finished his most famous publication "A History of the Species of Crocus" only a day before his death in 1847.

Crocus herbertii grows in wet meadows, on roadsides up to the Ulu-Dag national park, in clearings in pine forests alongside *Crocus biflorus* subsp. *pulchricolor* and sometimes even with *C. chrysanthus* which usually prefers drier spots. Our team went to the Ulu-Dag a little too early in the season, and therefore only saw the beginning of its blooming, but other travelers showed me pictures where its flowers cover the ground like a golden yellow carpet with some blue spots in between from flowers of *C. biflorus* subsp. *pulchricolor. Crocus herbertii* grows in places more accessible than those of *C. gargaricus* and is more widely distributed in gardens.

The two species are very similar, although *Crocus herbertii* has smaller corms and the corm tunics are finely fibrous reticulated only at the top. The corms form side-growing stolons. Despite its distinctly fibrous tunics, it is positioned in the phylogenetic tree alongside two beautiful blue-colored species—*C. leichtlinii* and *C. kerndorffiorum* from series *Biflori*—with membranous tunics.

In the garden, *Crocus herbertii* seems to be even hardier than *C. gargaricus*; at least I never lost it when it grew outside. In my first garden it is still growing near old apple trees where it has received no attention for at least twenty years. In the garden its stoloniferous habit is a great advantage. Once planted it slowly spreads occupying more and more new space. Of course, what is good for an ornamental garden is not nearly so good for nurseries that sell plants.

Crocus herbertii produces plenty of very small cormlets at the ends of its stolons. These are not easy to notice let alone to collect. Many of them remain unharvested and become a sort of a pleasant "weed" on nursery beds. The corms are very small, too, and even the smallest of them, in size rarely exceeding 5 mm in diameter, produces a perfect bloom. I always feel uncomfortable when dispatching these corms to my customers just because they are so small.

After blooming, *Crocus herbertii* likes drier conditions than does *C. gargaricus* and it doesn't suffer at all from summer rains. When grown in pots care must be taken not to overdry them during the summer rest.

Crocus sieheanus Barr ex B. L. Burtt

The name of *Crocus sieheanus* ($2n = 16$) commemorates W. Siehe who collected and introduced in cultivation many bulbous plants from Turkey. This one is distributed in central southern Turkey where it grows on open hillsides and sparse pine woodlands from 1200 to 2000 meters. In the phylogenetic tree it is placed alongside *C. gargaricus*.

The flowers are almost indistinguishable from those of *Crocus ancyrensis* and *C. chrysanthus*. It differs in having a bright red three-branched style that usually much exceeds the pale yellow anthers. Anthers can be with or without black basal lobes. The corm tunics, however, are very distinct. They are membranous at the apex, splitting into parallel or weekly reticulated fibers. The flowers are bright orange-yellow with a yellow throat and a yellow or purplish brown perianth tube.

Crocus sieheanus is very rarely cultivated and I have never had it in my collection, although several times I bought corms labeled as *C. sieheanus*. They turned out to be either *C. vitellinus* or *C. graveolens*; both are easy distinguishable by their many-branched style. In cultivation *C. sieheanus* requires a long and dry summer rest and is suitable only for pots.

Crocus graveolens Boissier & Reuter

PLATES 239–242

E. A. Bowles wrote of *Crocus graveolens* ($2n = 6$) the following:

> [I]n most seasons the first indication I receive of its being in flower is this pungent odour, which reaches my nose before the yellow of the flower catches my eye. . . . Dry specimens retain this scent for many years, and it even defiles the paper in which they have been pressed.

Although the specific name means "strong smelling," I have never noticed this trait even when I've put my nose into an open flower of the several different colored samples that I have. I can only say that each nose is different.

In the wild this species occurs in southern Turkey, northwestern Syria, and through Lebanon reaches northern Israel where it grows on limestone formations in stony and rocky places, often in scrubs and clearings of sparse woods, usually in rich red earth from 500 to 1600 meters. All samples growing in my collection originate from the northern part of its area, namely, Turkey where it is a very common plant.

The flower buds usually have a distinct "waist," but when they open widely in the bright sun the tepals curve outwards in the middle. The inside color of the flower is not very variable, being brighter or paler yellow, more often on the brighter side. The markings on the exterior of the sepals range from a grayish suffusion in paler forms to nicely brownish purple striped or, in the best forms, with a medium wide deep brown or purplish midzone which narrows to the tips of sepals. There are forms with only indistinct grayish stripes on the sepal outsides, and from the surroundings of Silifke come plants with very pale yellow flower segments which can be even described as straw-colored. My favorites are the brightest colored flowers with the most contrasting sepal backs. The flower segments are acute and the inner segments usually are slightly smaller than the outer. The style is divided into many yellow-orange branches which spread approximately at the tips of the yellow anthers. Very rarely anthers can be black. The leaves are gray-green. The corm tunics are membranous splitting into parallel strips or fibers at the base. In the wild the corms lie deep in the ground.

Crocus graveolens grows in localities which in summer are dry and hot, so it is suitable only for growing under cover where it performs quite well. It increases well by splitting and readily sets seed.

Crocus vitellinus Wahlenberg

PLATES 243–245

Crocus vitellinus ($2n = 8$) is very similar to *C. graveolens* but is reported as very sweetly scented. It is distributed in southern Turkey, western Syria, and Lebanon, where it can be found from sea level up to 1400 meters on limestone formations in open rocky places and sparse woods. The species name refers to the bright yellow-orange color of the Lebanese forms and means "egg yolk."

In the wild the flowering period stretches from late autumn (November) in most southerly distributed populations until April at higher elevations in the northern part of the area. *Crocus vitellinus* differs from *C. graveolens* by the shape of its flowers, which are slender and funnel-like but without a distinct waist, and by the flower segments, which are wider and usually rounded at the tips. The variation in flower color, however, is almost the same for both species, and I more and more tend to think that *C. graveolens* must be regarded only as a variety of *C. vitellinus* regardless of the different chromosome numbers. Both are very close neighbors on the phylogenetic tree, too.

The best, most deeply orange-colored form of *Crocus vitellinus* comes from Lebanon and usually blooms from the end of November into December, so it can also be listed among autumn-blooming species. Unfortunately, these bright forms are very difficult to preserve through dull and dark winter months. I lost my plants collected by Arnis Seisums in Syria after two or three seasons. The leaves are dark shiny green.

More common in collections are the easier-to-grow spring-blooming forms of *Crocus vitellinus* from southern Turkey, but they are paler in color. The most common form in cultivation comes from Dutch nurseries; it has medium yellow flowers with a grayish speckling over the back of the outer segments which becomes deeper at the base. Its throat is deeper yellow, even orange shaded. It increases fast by division of offsets but has never set seed for me.

I have a sample gathered as *Crocus sieheanus* at Findikpinar, southern Turkey (BATM-001), during a joint expedition with botanists from the Gothenburg Botanical Garden. When the flowers opened the following spring, we noticed our mistake: the flowers had a many-branched stigma. The color was a much brighter yellow, but the flower segments were much narrower and prominently rounded at the tips and they did not overlap when half-opened like in the Dutch sample. Although this locality is quite far from the earlier known area and coincides with the area of the closely related *C. graveolens*, the sample resembles *C. vitellinus* in every feature.

Like *Crocus graveolens*, *C. vitellinus* is a plant from dry and hot regions, so it can be grown only under cover. In summer it requires a dry and hot rest.

The Reticulated Blues

In this group I have combined blue-flowered species from series *Reticulati* with a couple of species from other series that have corm tunics similar to those of series *Reticulati*. Yellow-flowered species from series *Reticulati* have been reviewed in the previous chapter.

According to phylogenetic studies, the blue-colored species from series *Reticulati* are not closely allied and form two groups. One group comprises *Crocus veluchensis*, *C. dalmaticus*, *C. sieberi*, *C. atticus*, and *C. rujanensis*. The remaining species—*C. reticulatus*, *C. hittiticus*, and *C. abantensis*—are not very closely related and are positioned quite far from the first five species. For down-to-earth gardeners the important characteristic of these two groups is that their tunics are finer or more coarsely reticulated throughout and, in typical forms, the flowers are more or less blue- or violet-colored, although more or less distinct albinos are not rare among them.

Also included here is *Crocus etruscus* from series *Verni* because it has a reticulated tunic. The second outlier is *C. baytopiorum*, which Brian Mathew included in series *Verni* in his monograph but has subsequently separated it into a series of its own, namely, series *Baytopi*. Both of these species are absolute outliers and disrupt, to a certain extent, the natural grouping of this chapter as they belong to section *Crocus* in which the prophyll is present while all the other species, reviewed in this chapter, belong to Section *Nudiscapus*. I have to include them here to make their identification easier because of the distinctly reticulated tunics. Species with tunics reticulated only at the apex are discussed in the next chapter.

1. Anthers blackish maroon .... *C. hittiticus*
1. Anthers yellow
    2. Flowers white inside
        3. Outer flower segments acute or subacute, usually with three to five conspicuous stripes, open flowers look starry .... *C. reticulatus*
        3. Outer flower segments subacute to rounded, usually without stripes but with a purple staining or one central stripe or broad horizontal purple bands on outer segment outsides, flowers globular .... *C. sieberi*
        3. Flowers without stripes or staining on outside .... white forms of *C. veluchensis*
    2. Flowers lilac or bluish inside
        4. Throat white or pale yellow
            5. Bract only present, clearly visible sheathing perianth tube, prophyll present
                6. Flowers lilac, throat pale yellow .... *C. etruscus*
                6. Flowers of very unusual color, nearly greenish blue, throat of the same color or whitish .... *C. baytopiorum*
            5. Bract and bracteole present, prophyll absent
                7. Outer segments usually striped .... *C. reticulatus*
                7. Outer segments without stripes

8. Flowers strictly funnel-shaped .... *C. veluchensis*
8. Flowers with pronounced waist .... *C. atticus* subsp. *sublimis*

4. Throat yellow
   9. Leaves 1 mm wide or less, flower segments usually without veins . . . *C. abantensis*
   9. Leaves 1–2 mm wide, flower segments usually buff, yellowish gold, or silvery on the outside, often with dark veins .... *C. dalmaticus*
   9. Leaf width 2 mm or more, flowers lilac or violet
      10. Corm tunics fairly coarsely reticulated
         11. Corm tunics with long brown persistent neck of old sheathing leaves, bract and bracteole more or less equal in length, leaves 2–6 mm wide .... *C. atticus* subsp. *atticus*
         11. Corm tunics without persistent neck at apex, bract and bracteole unequal in length, leaves 2–3 mm wide .... *C. rujanensis*
      10. Corm tunics fairly finely reticulated without persistent neck at apex
         12. Throat of perianth pubescent .... *C. atticus* subsp. *sublimis*
         12. Throat of perianth glabrous .... *C. atticus* subsp. *nivalis*

Crocus sieberi J. Gay

PLATES 246–249

Crocus sieberi (2*n* = 22) *sensu lato*, at least in the range of some of its subspecies (here regarded as *C. atticus*), is the most well-known crocus in gardens from this grouping of species. It is named after Franz Sieber, who collected plants on Crete which J. E. Gay used to describe this species. Thus, the Cretan plants must be regarded as the type to which the name of *C. sieberi* must be applied.

The species is found on rocky slopes and grazed mountain grassland from 1500 to 2700 meters where it flowers from March to June depending on the season and elevation. It is a Cretan endemic confined to three main massifs and Kedros.

Crocus sieberi is one of the most attractive crocuses with extremely variable flowers: white on the inside with a bright yellow or even orange throat and veined or suffused with purple to blackish violet on the outside in various patterns. Rarely the plants can be pure white. At the other end of the spectrum are the distinctly bicolored forms with outer segments having a deep purple back and inner ones being pure white. The purple coloration on the backs of the outer tepals can be in form of a suffusion of various lengths, like a central stripe or a purple blotch at the tip or both on the base and at the tip. Some of the most spectacular forms have purple outsides crossed with a narrower or wider horizontal white stripe near the tips. If it is common in *Crocus* that flowers with a dark tepal back are edged white, among forms of *C. sieberi* are some in which the white sepals have a narrow purple-feathered rim (Plate 247).

I obtained my first corm of this beauty from a Czech friend who collected it on Crete as an unnamed *Crocus*. It turned out to be a marvelous pure white form with only slight purple shading on the base of the outer segments. It increased well by division of offsets and was self-fertile, yielding plenty of seed. This original clone was so beautiful and abundantly blooming that I named it **'Cretan Snow'** (Plate 248). In color it is somewhat similar to the well-known 'Bowles' White' (see *C. atticus*), but is smaller in size, more starry, and with a lighter yellow throat. Its seedlings are very variable both in size and in the amount of purple on the flower backs. Many of these are worth selecting and evaluating as potential candidates for cultivar names. One of the best is a form with entirely purple sepal backs and a narrow white midzone like a small island of white ice in a sea of dark water.

Other variable forms that I got from different collectors all had generally much darker colored sepal backs. An especially beautiful form was sent by John Fielding from the United Kingdom. In color it was somewhat similar to the cultivar **'Hubert Edelsten'** but with a wider and more distinct white band crossing the sepals at the tip and of a better flower shape. This clone's greatest advantage was its fertility as 'Hubert Edelsten' is a sterile hybrid between *Crocus sieberi* and *C. atticus* from mainland Greece, although on the Internet are reports about occasional seeds when it was grown alongside 'Tricolor'.

The only fault of *Crocus sieberi* is its insufficient hardiness. When planted outside in my garden, it has a very short lifespan and is killed in the first unfavorable winter. I have never managed to keep it longer than two to three seasons outside. Potted plants in the greenhouse, however, have never suffered from the winter vagaries. 'Hubert Edelsten' is much hardier, though I lost it, too, after a very harsh winter; fortunately, it can be replaced thanks to Dutch nurseries, and I now plant a small stock of it also in the greenhouse.

The famous Dutch breeder of crocuses and other small bulbs, Willem van Eeden, selected a mutation of 'Hubert Edelsten' and named it **'George'** (Plate 249) in honor of the renowned Russian iris specialist George Rodionenko. It is similar to its parent in beauty but has larger and more rounded flowers and a narrower white zone, and it flowers a few days later. Its greatest advantage is the enhanced hardiness. I have lost 'Hubert Edelsten' several times but never 'George'.

In 1994 Alan Edwards spotted an unexpected yellow seedling among a group of plants raised from seed of *Crocus sieberi*. He considered the new form to be on the same level with 'Bowles' White', but with the white surface color replaced by two tones of yellow. Edwards later named it **'Midas Touch'**. The seeds had come from cultivated plants of *C. sieberi* grown by Lyn Bezzant in Scotland. These plants undoubtedly were a hybrid with an unknown species, perhaps *C. chrysanthus* or *C. cvijicii*. The latter seems more probable because like *C. sieberi* it also belongs to series *Reticulati*.

I recently tried to cross a late white-blooming form of *Crocus sieberi* with the pollen of a golden yellow *C. cvijicii* and got some seeds. Because I don't isolate hand-pollinated flowers, I won't know whether I succeeded until the seedlings bloom.

Cretan *Crocus sieberi* is the most difficult plant of *C. sieberi–C. atticus* complex in cultivation. Because it is less hardy than its allies from mainland Greece, it is no wonder that it is still quite rare in cultivation and seldom seen in collections. Despite this, it is an easy plant to grow in pots, provided you remember that, unlike its mainland allies, this one likes a somewhat drier and warmer summer rest.

Greek mainland allies of *Crocus sieberi* generally have larger lilac-blue flowers. I like to regard them as different species, as they are so different from the smaller purple-colored Cretan plants. The best way to handle this would be to regard the Cretan plants as *C. sieberi* but those from the Greek mainland as *C. atticus* divided (or not?) into three subspecies. Alan Edwards (and David Stephens agrees with him) questions the significance of the presence or absence of throat hairs in various samples from many different collections of the mainland *C. sieberi* allies as the basis for splitting them into subspecies.

Crocus atticus (Boissier & Orphanides) Orphanides

PLATES 250 & 251

Crocus atticus ($2n = 22$) is a more robust plant with up to 6 mm wide leaves and entirely lilac-blue or violet flowers which have a bright yellow or orange throat. The corms are coarsely reticulated with a persistent neck of old sheathing leaves. The species occurs in Attica, southern Euboea, and on the Andros Island.

The cultivar now grown in the gardens under the name **'Bowles' White'** (Plate 251) most likely belongs to *Crocus atticus*. E. A. Bowles wrote about it: "In 1923 I found two pure white youngsters among my seedlings, after thirty years of hopeful expectations. The better of these has increased freely and is the best white and orange spring crocus I know." It received several awards but for some time was thought to be lost in cultivation.

In the 1980s, during my first trip to the Netherlands, Willem van Eeden presented me with 10 corms of this exceptionally beautiful crocus. It is a good increaser and vigorous but doesn't belong to the hardiest forms of *Crocus atticus*. Several times I lost corms planted outside due to very severe winters. I have never obtained seeds from this clone. Unfortunately several commercial stocks are now virus infected.

Some gardeners reported that their clone is fertile (the International Register lists another cultivar from E. A. Bowles named 'Albus') and that it would be better to name it 'Albus'. I don't know who is right. Perhaps both variants are still cultivated. I think that the contribution of E. A. Bowles to our present-day knowledge of crocuses is so substantial that his memory must be preserved in the name of this undoubtedly marvelous crocus.

There are several cultivars in the trade offered as related to *Crocus atticus* subsp. *atticus*. The most popular are cultivars **'Firefly'** and **'Violet Queen'**. They all are good growers and quickly increase by splitting and at least the commercial form offered as **'Atticus'** sets abundant seed. These forms are hardy and suitable for gardens, only suffering in very unfavorable winters.

Crocus atticus* subsp. *sublimis (Herbert) Rukšāns, comb. nov.

PLATES 252–254

Crocus atticus subsp. *sublimis* (Herbert) Rukšāns, comb. nov. *C. sublimis* Herbert in Bot. Reg. 31: misc. 81 (1845). Syn. *C. maudii* Maly, *C. athous* Bornmüller, *C. sieberi* subsp. *sublimis* (Herbert) B. Mathew.

Crocus atticus subsp. *sublimis* is more widespread than subsp. *atticus* and can be found from northern Peloponnesus northwards to Albania, southern Yugoslavia, and southern Bulgaria. It grows in alpine grassland or woodlands where it blooms near snow melt from 1500 to 2300 meters. It is somewhat paler in color than the type subspecies, although very bright colored forms can be found as well. The throat usually is of paler yellow color.

One of the best forms of subsp. *sublimis*, which is very widely grown in the gardens, was found on Mounts Chelmos and Ziria. The flowers are bright lilac with a very bright orange throat that is well separated from the lilac part of a flower by a wide white zone. This white zone is visible even on the outside surface of closed flowers. This variant was named **forma *tricolor*** by B. L. Burtt in 1949, and one of the most vigorous selections was later named **'Tricolor'** (Plate 253). There is a great variation among the wild forms of f. *tricolor*. Brian Mathew reported one with a fourth color on the tepals—a very deep purple blotch at the tips of the segments.

In the late 1980s I bought several corms of the then-new species *Crocus robertianus* from the renowned Dutch bulb specialist and nurseryman Michael Hoog. Instead of the expected autumn-blooming species, it turned out to be a very nice form of the spring-blooming *C. atticus*. The purchased corms had been collected in a locality where *C. robertianus* grew together with *C. atticus* subsp. *sublimis*, and as it was impossible to distinguish these two species by corm tunics I was not very disappointed because the next year they were generously replaced with bulbs of the true species. This form of *C. atticus* is very vigorous and absolutely hardy, and increases readily. The outside of the flower segment is somewhat silvery lilac, densely covered with narrow darker violet stripes, and the throat of the flower is of the same bright orange shade as in the best forms of *C. sieberi* and *C. atticus*. In flower color this form somewhat resembles some specimens of *C. dalmaticus*, but the latter has narrower leaves. A bit confusing are the very coarsely reticulated corm tunics which in typical subsp. *sublimis* usually are finer. After the death of Michael Hoog I decided to name this clone in his honor **'Michael Hoog's Memory'** (Plate 254).

I have a representative sample of *Crocus atticus* subsp. *sublimis* from Mount Parnassus in which the flower throat is variably colored. In most specimens the throat is quite pale, far from the bright yellow usually associated with *C. sieberi* and *C. atticus*; in some of them the throat is completely devoid of the yellow pigmentation (Plate 252). Brian Mathew thinks that these most possibly are hybrids between *C. veluchensis* and *C. atticus* (*C. sieberi* according to B. Mathew). Alan Edwards reports that in some places hybrid swarms overwhelmingly dominate. Hybrids have the "waisted" shape and strongly marked foliage of *C. atticus*, are often bicolored in some degree, but lack any yellow in the throat as is typical of *C. veluchensis*, or the throat is only slightly creamy.

Subspecies *sublimis* is the hardiest and most vigorous variant of *Crocus sieberi sensu lato*. It grows well in the open garden as well as in pots. 'Tricolor' can suffer in hardiest winters but 'Michael Hoog's Memory' and a form from Mount Parnassus survived even in the extremely unfavorable winter of 2005–06 and suffered no damage.

In his monograph Brian Mathew mentions an interesting crocus collected by him near Langadia in northern Peloponnesus that was supposed to be *Crocus cancellatus* because of its very coarsely reticulated tunic and narrow leaves, but several of the corms bloomed in spring. It has a number of features separating it from the typical *C. atticus*. It grows in screelike conditions while the neighboring subsp. *sublimis* occurs in damp wooded areas. Probably this is another not-yet-described subspecies or even a species.

Crocus atticus subsp. *nivalis* (Bory & Chaubard) Rukšāns, comb. nov.

PLATE 255

> *Crocus atticus* subsp. *nivalis* (Bory & Chaubard) Rukšāns, comb. nov. *C. nivalis* Bory & Chaubard. Exped. Sci. Moree 3, 2: 21, t. 2, fig. I (1832). Syn. *C. sieberi* subsp. *nivalis* (Bory & Chaubard) B. Mathew.

The third subspecies, *Crocus atticus* subsp. *nivalis*, is distributed in southern Peloponnesus in the Taygetos Mountains where it flowers near snow melt in mountain grassland at 1100 to 2800 meters. It has a finely reticulated tunic and a glabrous throat.

I have only one sample of subsp. *nivalis* and it came from John Fielding. It was collected slightly beyond the traditional area of the subspecies, but its identification was confirmed by Brian Mathew. It is rarely cultivated, and I have tried it only in the greenhouse. My plants are somewhat lighter in color but with a bright yellow throat.

There are also forms with very bright lilac blooms, in which the small bright orange throat is edged by a narrow white zone. The variation is very spectacular though not as distinct as in 'Tricolor'.

Crocus rujanensis Randjelović & D. A. Hill

PLATE 256

Crocus rujanensis (2*n* = ?) was found in March 1984 on slopes of the Rujan Planina in southern Serbia between 350 and 750 meters. It was compared with *C. dalmaticus* and *C. atticus* subsp. *sublimis* (as *C. sublimis*) and found to be different in a combination of characteristics rather than in one distinct feature. It is spring blooming, has light to dark purple flowers with a yellow throat, its bract and bracteole are unequal in length, and the perianth tube is purplish yellow in the upper part. In the other two species, the bract and bracteole are equal in length and the perianth tube is purple in the upper part. *Crocus rujanensis* grows at lower elevations and in less mountainous places.

I have several samples of this species collected by various travelers and differing in color. The stock grown from seeds collected by Jim and Jenny Archibald at the *locus classicus* is somewhat darker in color with a very bright orange throat in some specimens surrounded by a narrower or wider diffused white zone. At the tips on the outside of the flower segments is a darker purple, though not very contrasting V-shape mark of the type so common in *Crocus heuffelianus* but less prominent. Such marks are absent in other samples which in general are lighter in color but retain the same overall appearance. In my samples the yellow, orange, or even red style branches well exceed the tips of anthers. I have seen similarly positioned style branches among some forms of *C. atticus*, too.

Crocus rujanensis is not difficult in cultivation. In my garden it does better in a greenhouse than outside. It seems our winters are too harsh for it. Although I have never lost corms that were planted outside, their blooming becomes much poorer after a particularly cold season. I will keep my stocks of this very beautiful plant, although I don't like species that can be distinguished only by a "combination of characteristics." They look too intermediate to other species used for comparisons. It was not easy to find features that could be used in the proposed key to identify this species, and the possibilities for misidentification are numerous. David Stephens (1998, p. 357) comments on it: "*C. rujanensis* from Serbia is to all intents and purposes a disjunct form of *C. sieberi* subsp. *sublimis*, though it is distinct enough in its own right to keep the synonym written on the reverse of label."

Crocus dalmaticus Visiani

PLATE 257

The name *Crocus dalmaticus* ($2n = 24$) is well known in the trade but every time I buy Dutch-grown corms under this, I receive plants with much wider leaves that are indistinguishable from *C. atticus* 'Firefly'. The true *C. dalmaticus* occurs in southwestern Yugoslavia, in the now-independent states of Croatia, Bosnia and Herzegovina, and in Montenegro and northern Albania, stretching along the Adriatic coast. It grows on limestone formations in rocky grassland and deciduous scrub from 300 to 2000 meters.

In general *Crocus dalmaticus* is quite similar to some forms of *C. atticus* but has outer segments with a paler backside, usually buff or silvery shaded, and has leaves that are narrower. The style in *C. dalmaticus* is rather slender and deeply trilobed whereas in *C. atticus* it is a substantial structure with a frilled appearance. Regardless of the different chromosome numbers, Brian Mathew reports that he has succeeded in producing hybrids between them.

The flower segments of the best forms have a bright lilac interior and a yellowish or buff exterior, especially when the flowers are half-open. My stock, originally collected at Petrovac in Montenegro, has very variably colored backs of the flower segments. Some have slightly buff-toned sepal outsides and some are very similar in color to my selection from *Crocus atticus* subsp. *sublimis* 'Michael Hoog's Memory', only the throat in *C. dalmaticus* is lighter yellow. According to Brian Mathew, there are populations with flowers from almost white to deep lilac-purple, with markings on the outside in varying colors and intensity.

This plant has been long known in cultivation. The first records of it in cultivation are from 1840. It is not easy to judge how widely the true species is now grown because of 'Firefly' which is regularly incorrectly named as *Crocus dalmaticus*. But the true *C. dalmaticus* is an easy plant and grows well in the open garden and in the greenhouse. It increases fast by division of offsets and sets abundant seed. It doesn't like too dry conditions in summer, so the pots must be taken outside during the rest.

Crocus veluchensis Herbert

PLATES 4, 11, 258 & 259

Crocus veluchensis (2*n* = 18, 19, 20, 22, 26) is among the latest-blooming crocuses in my collection. It was named after the Greek mountain of Veluchi (now renamed Timfristos) and occurs in central and northern Greece, Albania, southern Yugoslavia, and Bulgaria, where it grows in alpine meadows and woods from 950 to 2600 meters.

This species is extremely variable in size and flower color as well as in suitability for cultivation in the gardens. Brian Mathew reports that some forms are difficult. E. A. Bowles had a similar experience. All the specimens that I have acquired have proved to be good growers.

Typical forms of the species have deep purplish lilac flowers though white ones are not rare. This species is easily separable from *Crocus atticus*, especially from subsp. *sublimis*, by the more slender flowers and the lack of any yellow in the throat. On the whole the flowers are larger, although small forms are also reported. I grow several selections of it, mostly with large purest white or brightest purple flowers. One form has very slightly bluish shaded flower segments, not exactly albinos but not far from it. Some of my selections have smaller flowers. The color of one form very closely resembles some of the paler purplish forms of *C. heuffelianus* with the deep purple blotch at the tips of the tepals as short narrow stripes instead of a confluent blotch; this form belongs to the small-flowered type of *C. veluchensis*.

Some forms resemble *Crocus vernus* in flower shape and color, but differ in the corm tunic, which is finely reticulated in *C. veluchensis* and parallel or only slightly reticulated in *C. vernus*. Sometimes it is not easy to separate these two species by the corm alone but, fortunately, *C. veluchensis* has one very special feature which enables easy separation from all the other similar crocuses—it starts forming the new roots before the end of vegetation and before the roots of the previous season have died off. Thus, the best time for replanting the corm is immediately after lifting it. I believe that this feature alone is the reason why this marvelous and beautiful crocus is not offered by the big nurseries. In fact, it doesn't have to be an obstacle if the lifted corms are stored in a thin plastic bag filled with peat moss. Despite the presence of roots, the corm endures transportation well and restarts growing when planted in the new place. Sometimes I have received corms with shriveled new roots, and they grew on just like the corms from the plastic bags, only the blooming the first spring was weaker.

Crocus veluchensis is completely hardy in the garden and I have never had problems with it when planted outside even after the coldest and most unfavorable winters. Possibly it is due to the late blooming and late awakening from the winter dormancy. Its blooming starts when most of other crocuses are over. *Crocus veluchensis* readily sets seed, which germinates even when sown on open garden beds; the first blooms appear starting from the third year. The seedlings vary greatly in color.

The species can hybridize with *Crocus cvijicii* (see discussion in the previous chapter). Brian Mathew and Alan Edwards report on possible hybridizing with *C. atticus* subsp. *sublimis*. When growing corms in pots, it is important to carefully check the moisture level of the soil mix. In summer *C. veluchensis* likes a cooler position and benefits from being watered during the dry period.

Crocus reticulatus Steven ex Adam

PLATES 260 & 261

Phylogenetic research shows that *Crocus reticulatus*, *C. hittiticus*, and *C. abantensis*, which were placed by Brian Mathew in series *Reticulati* with species from the *C. sieberi* group, actually are quite distant relatives and are more close to species from the *C. biflorus* group. The same study placed both subspecies of *C. reticulatus* far from each other. For these reasons I am following the earlier viewpoint of Brian Mathew, who at first regarded *C. hittiticus* as a separate species and only later reduced its status to a subspecies of *C. reticulatus*.

Crocus reticulatus ($2n$ = 10, 12, 14) was one of the first wild species which I started to grow in my early youth. Its name characterizes the coarsely reticulated corm tunic. This very widespread species occurs in the wild from northern Italy through the former Yugoslavia, Bulgaria, Romania, Hungary, Ukraine, Crimea, and the Caucasus, reaching Turkey. It grows in dryish grassland and on bare rocky slopes, sometimes in open woods, from sea level up to 2100 meters. In Georgia I collected it in very dense turf. It is not surprising to find great variation in such a large area. The shape of the tepals is very special: it seems that *C. reticulatus* (and its relative *C. hittiticus*) is the only species in which the outer flower segments are narrower than the inner (whereas in all the other species it is just the opposite).

My first corms of *Crocus reticulatus* were gathered in Moldavia (formerly Bessarabia) near windbreaks partitioning cultivated fields and in roadside plantings where they were very abundant locally—as many as 50 specimens per square meter (Plate 260). The form produced by these corms turned out to have a very beautiful, bright flower color and is an excellent grower and multiplier for me. Later I was very surprised to learn that it has never been a particularly well-known plant in horticultural circles, "probably because it lacks vigor in most northern gardens and in addition is not very showy" (B. Mathew). My experience is contrary to this statement. At that time I was growing my first corms, I was familiar only with the population from Moldavia. Its flowers are nicely deep purple-striped over the outside of flower segments; in full sun the flowers open somewhat star-shaped and are bright lilac. They are elevated at a distance from the ground by quite long tubes. The flower throat is light yellow. This population is fairly uniform in color, only several plants are lighter lilac in shade, and it readily sets seed and multiplies vegetatively. Until the most recent very unfavorable winter, I never experienced any problems with it growing outside.

Now I have many samples covering the species's range from Italy, Slovenia, the Azov steppe in Ukraine, the Stavropol district in the northern Caucasus, and Turkey. Although all the samples look alike when the buds are closed, this changes when the flowers open. The most variable population is near Pjatigorsk in the Stavropol district. There the inside of the flowers varies greatly from almost white to deep lilac and the most interesting feature is that the plants with a white inside have a yellowish throat, but plants with a lilac inside have a white or even grayish throat. Some specimens from Slovenia have a nice yellowish buff toned base color on the outside of sepals. According to Brian Mathew, the external markings are nearly absent in some colonies from central northern Turkey.

Plants of *Crocus reticulatus* collected by my Czech friend near Gülek Pass (province Içel), Turkey, are very special (Plate 261). Their base color is almost white with light lilac shading and on the outside they are covered with narrow though closely spaced deeper colored stripes. It is the lightest form known to me, but Brian Mathew states that there can be specimens with fine speckling in the form of dots and dashes. At Gülek Pass *C. reticulatus* grows with *C. hittiticus*; the latter always has darker-colored flowers but with a similar color pattern on the sepal backs, only the anthers have black connectives. Possibly this is the place where the divergence of both species began. More to the south stretches the area of the black-anthered ally.

Crocus reticulatus hybridizes easily with yellow *C. angustifolius*, when the two are planted near each other. Hybrids were reviewed in the previous chapter along with *C. angustifolius*.

Brian Mathew recommends growing this species in pots and giving it a dry summer rest. I provide such treatment for the Turkish samples, but those from Ukraine and the Caucasus can be grown outside as they don't suffer much from summer rains. They perfectly set seed and some are good increasers by splitting. I do keep a few plants from each sample potted in the greenhouse. In any case *Crocus reticulatus* is a beautiful plant with a strong honey perfume, and I recommend it for every crocus collection.

Crocus hittiticus T. Baytop & B. Mathew

PLATE 262

Crocus hittiticus ($2n = 10$) is much weaker in cultivation than *C. reticulatus* and is not suitable for the open garden. It was described as a species by Turhan Baytop and Brian Mathew in 1975, but later reduced in status in Brian's monograph as *C. reticulatus* subsp. *hittiticus*. It is known from the southern slopes of the Cilician Taurus in the province of Içel, Turkey, where it grows on limestone rocks in woodland or in forest clearings from 750 to 1400 meters.

In flower color it is similar to *Crocus reticulatus*, but more often the exterior of the flower segment is speckled. Most spectacular are its large shining black or dark purple anthers. I especially like a form collected by Jim Archibald, Arnis Seisums, and Norman Stevens (SASA-022) with distinctly striped backs of the segments and very large black anthers, although other forms with more diffused sepal outsides are not less beautiful.

I obtained my first samples of *Crocus hittiticus* before I had a greenhouse. They survived for a number of years but flowered reluctantly. The substance of the flower segments was so poor that every spring the flowers suffered from swings in the weather. Eventually, I lost all my plants. When I planted new corms in the greenhouse, they grew far better. They flourished with abundant flowering, increased at a good rate, and started to set seed. *Crocus hittiticus* requires dry and warm summer conditions.

Crocus abantensis T. Baytop & B. Mathew

PLATES 263–265

Crocus abantensis ($2n = 16$) is distinct from other species with reticulated corms and blue flowers. It is distributed at high elevations near Lake Abant in northwestern Turkey where it blooms on mountain meadows near melting snow between prostrate junipers and in open spots in low pine forests from 1100 to 1350 meters. It grows together with the golden yellow *C. ancyrensis* and in the same area are distributed (although I haven't seen them growing together) the similarly colored *C. biflorus* subsp. *pulchricolor* and another golden species, *C. olivieri* subsp. *olivieri*. Most likely it was the overall resemblance to *C. biflorus* that caused *C. abantensis* to be overlooked for such a long time in this well-visited area, but any doubts about their identity disappear when you see a corm because the two have very different corm tunics.

This species blooms when the weather conditions can be very fickle. In spring 2007 I visited the heights over Lake Abant on the 11 March. *Crocus abantensis* and the neighboring *C. ancyrensis* were in full flower, but the slopes were still covered with plenty of snow. Two weeks later on my way back I again visited these places and everything was covered by deep snow and I found only few plants of *C. abantensis* pushing tips of flower buds through the snow blanket. In 2008 our team was at the site on 20 March when flowering was just beginning. In some spots *C. abantensis* dominated, in other places only *C. ancyrensis* grew; mixed groups were not a rare occurrence. Despite their differing chromosome numbers, the two species sometimes hybridize. Their hybrid with lilac-flushed yellow flowers was described as *C.* ×*paulineae* (see "Yellow Fever" and Plates 266 and 306).

Flower color of *Crocus abantensis* can vary widely from mostly blue to rarely white, and at worst is a dull slaty shade. I received a deep lilac specimen from the Gothenburg Botanical Garden. Another especially beautiful form has a very light base color, densely covered with nice light blue stripes. I named a striped form **'Azkaban's Escapee'** (Plate 265) from the Harry Potter novels. Its pattern somewhat resembles vertically striped prisoner's pajamas. The flower throat is invariably deep yellow; the three yellow or bright orange style branches end slightly below or approximately at the same height as the yellow anthers, rarely slightly surpassing their tips. The corm tunics are finely fibrous.

Crocus abantensis is easily distinguishable from its closest relative *C. reticulatus* by the more rounded flower shape, the blue color (*C. reticulatus* is never blue), and the much finer reticulated tunics. It is a very narrow endemic known only from the type locality although there it occurs very abundantly. The species increases well by vegetative means.

In the garden *Crocus abantensis* is a very hardy and easy-to-grow plant. I grow it mostly outside; in pots I keep only its color variants to protect them from rodents rather than the whims of weather. In the winter of 2005–06, my plantings in the open garden seriously suffered because *C. abantensis* is an early blooming species and was in full bloom when a very hard black frost set in. Fortunately, such conditions are very rare. It readily sets seed and self-seeds abundantly on the paths between nursery beds, so with me it has become almost a weed. It increases well vegetatively. Since *C. abantensis* comes from a cold region with lots of snow in winter and a not very dry or hot summer, it doesn't require summer baking. When grown in pots, it must be placed outside during the summer.

Crocus etruscus Parlatore

PLATE 267

Crocus etruscus ($2n = 8$) is included here only because of its coarsely reticulated tunic, but it belongs to a different section where prophyll is present. Its closest relative is *C. albiflorus* treated in the next chapter. In the wild it occurs in northern and northwestern Italy where it grows in mixed deciduous woods and on fields at relatively low elevations from 300 to 600 meters. There are some records from Corsica where it was found at much higher elevations.

Crocus etruscus in general is a lighter or darker pale lilac with creamy, silvery gray or pale buff on the exterior of the flower segment. There are reports about forms with distinct stripes or feathering on the exterior of the outer segments, but I have never seen any myself. In my specimens there are only short stripes at the base and the tepals are rather narrow, although the clone distributed under the name 'Rosalind' has much wider flower segments. The Corsican plants are more buff, even yellowish on the outside. The throat is very small, hidden deep between the tepals, and pale yellow; at first glance it sometimes creates an impression that the inside of the segments is pale lilac throughout and even whitish at the base, but more careful inspection confirms the presence of the yellow. The filaments are pale yellow, the anthers yellow. The three-branched style is more or less at the same level as the tips of anthers.

Two cultivars are common in cultivation. **'Rosalind'** is very pale colored with almost no stripes on the backs of tepals, and only three short lines at the very bottom of the outer segments. **'Zwanenburg'** is a much darker lilac than the typical samples, with a very narrow median stripe almost reaching the tips of outer segments and some shorter lateral stripes almost converging at the base and running down the tube.

Crocus etruscus is a very easy plant to grow in the open garden. Its somewhat slender stature and generally paler color, however, are not so remarkable. Thus this species is overshadowed by its much brighter relatives. It is easy to grow in pots and doesn't need a dry and hot summer rest. Although this species isn't a favorite of mine, it is good for plantings nears trees or shrubs as it grows easily outside and can even naturalize.

Crocus baytopiorum B. Mathew

PLATE 268

Crocus baytopiorum (2*n* = 28) has a distinctive flower color and is named in honor of professors Turhan and Asuman Baytop who found many interesting crocuses in Turkey. The species is distributed in southwestern Turkey where it grows on limestone screes and in rocky gullies in sparse coniferous woods from 1300 to 2700 meters.

The species is very outstanding, especially for its most unusual color. It is somewhat variable blue but of a very special shade that allows it to immediately distinguished with absolute certainty from all the other "blues" which as a rule are more to the lilac or purple end of the spectrum. Because of its morphological features, it was originally included in series *Verni*, but Brian Mathew clearly noted that it has no close relatives. Later phylogenetic studies confirmed that it stands very far from other species in series *Verni* being much closer to series *Crocus* but distinct enough to be separated in its own series *Baytopi*.

In its best forms *Crocus baytopiorum* is one of the most beautiful spring crocuses and usually discernable from a great distance just because of the unusual color. Its flowers are pale blue with darker veining and a white, grayish, or pale blue throat. At the outside base it is darker and more greenish in shade, a little flushing downwards on the upper part of the tube. The filaments are pale yellow or white, the anthers light yellow. The style is generally yellow or orange, divided into three branches, each clavately expanding at the apex, in most cases below the tips of anthers. I have some specimens in which the style branches are white and quite significantly surpass the anthers.

In 2005 our team explored the bulbous flora some 30 kilometers from Denizli. I collected some corms identified on the spot as *Crocus cancellatus* subsp. *mazziaricus* because of the shape of the tunic and the locality where they were gathered. In autumn a few of them bloomed confirming my identification. They were very variable with pointed and rounded flower segments and a pure white throat with deep purple stripes in the throat in the best forms. I was very surprised when the same pot bloomed again in spring, this time producing an exceptionally beautiful form of *Crocus baytopiorum*. All my previous samples acquired from various sources were more bluish, but the newcomer was much more greenish and distinctly green at the outer base, especially before opening. It is practically impossible to separate both species by the corm tunics which are very coarsely reticulated and somewhat bristly at the apex.

In gardens *Crocus baytopiorum* is a fairly vigorous plant and I have never lost it, although each spring when the snow has melted, I search for it with a trembling heart to see if it has pulled through the winter. It has always greeted me with its unusual flowers. It even sets seed if the weather is sunny and warm at the blooming time. When I built the greenhouse I moved all my corms of *C. baytopiorum* to pots—for peace of mind. In pots under cover it grows better, flowers plentifully, and readily sets seed every season.

Other Blue-White Species

This very artificial group comprises all the species with blue and white flowers for which I didn't find a place in other groups. All the members of this group bloom in spring, although two species reviewed among autumnal bloomers are included in the key: *Crocus cambessedesii*, which usually flowers all winter until spring, and *C. laevigatus*, which has some forms that bloom in spring, including very late spring. *Crocus paschei* is reviewed with annulate crocuses but is included in the key for this chapter, too. Variation in flower color in most of the species reviewed in this chapter is very great and white-colored forms are very common.

1. Flowers pure white or white with gray or purplish suffusion outside
   2. Segment outside unmarked or sometimes stained blue or brown at base, style with three branches, branches sometimes lobed or expanded at apex
      3. Throat yellow .... *C. malyi*
      3. Throat white or bluish
         4. Flower segments 3–5.5 cm long, style much exceeding anthers .... *C. vernus*
         4. Flower segments 1.5–3 cm long, style much shorter than anthers ... *C. albiflorus*
   2. Outer segment outside suffused and spotted with gray or purplish
      5. Style with 6 branches .... *C. candidus*
      5. Style with many branches
         6. Tunic with long persistent neck .... *C. antalyensis*
            7. Stigmatic branches yellow .... subsp. *antalyensis*
            7. Stigmatic branches white or whitish .... subsp. *striatus*
         6. Tunic bristlelike at apex .... *C. boulosii*
   2. Outer segments with dark purple or bluish blotch at the tip of flower segments .... *C. heuffelianus*
1. Flowers white inside but the outside blue or purplish or white with purple stripes
   8. Throat yellow, style with more than three distinct branches
      9. Corm tunic with long persistent neck, flowers pale lilac-blue to deep lilac .... *C. antalyensis*
      9. Corm tunic smooth and coriaceous, flowers usually with 1–3 purple to deep violet stripes .... *C. laevigatus*
   8. Throat white, style three-branched, branches sometimes lobed or expanded at apex
      10. Flower segments 3–5.5 cm long, style much exceeding anthers .... *C. vernus*
      10. Flower segments up to 3 cm long
         11. Style usually much shorter than anthers, flower segments slender, oblanceolate or subacute, 1.5–3 cm long, blue or purple striped .... *C. albiflorus*
         11. Style branches +/- at the level of tips of anthers, segments obtuse (rounded), up to 2 cm long, very dark purple, even blackish striped .... *C. cambessedesii*

1. Flowers lilac inside
 12. Throat yellow
  13. Style with more than three distinct branches
   14. Anthers white, corm tunics coriaceous .... *C. laevigatus*
   14. Anthers pale to deep yellow
    15. Throat yellow without white zone, corm tunics with long neck... *C. antalyensis*
    15. Throat with a distinct white zone above the yellow, corm tunics papery without neck .... *C. paschei*
  13. Style three-branched, branches sometimes lobed or expanded at apex
   16. Leaves usually two, rarely three, perianth outside without stripes ..... *C. kosaninii*
   16. Leaves three to five and more, flowers usually striped darker on outside
    17. Leaves gray-green with at least one prominent rib in each groove on the underside .... *C. versicolor*
    17. Leaves deep green without prominent ribs in the grooves beneath
     18. Bracteole present, filaments 6–9 mm long, anthers 12–21 mm long, style branches deep orange .... *C. imperati*
     18. Bracteole absent, filaments 3–5 mm long, anthers 8–12 mm long, style branches yellow or orange .... *C. suaveolens*
 12. Throat not yellow
  19. Flowers uniformly colored externally
   20. Flower tube of the same color as flower segments, leaves usually 4–6 mm wide
    21. Flower segments 3–5.5 cm long, style much exceeding anthers ..... *C. vernus*
    21. Flower segments 1.5–3 cm long, style much shorter than anthers .... *C. albiflorus*
   20. Flower tube white or very pale, leaves 2–3 mm wide .... *C. tommasinianus*
  19. Flower segments without stripes but with a dark V-shaped blotch at tips
   22. Stigmatic branches mostly approximately at the same level as tips of anthers, flower throat nude .... *C. heuffelianus* subsp. *heuffelianus*
   22. Stigmatic branches well exceeding tips of anthers, flower throat hairy .... *C. heuffelianus* subsp. *scepusiensis*
  19. Flower segments usually striped outside
   23. Corm tunics parallelly fibrous, finely netted towards the apex .... *C. corsicus*
   23. Corm tunics parallelly fibrous throughout
    24. Flower segments 2–2.7 cm long, leaves markedly deep green, 0.5–1.5 mm wide .... *C. minimus*
    24. Flower segments 2.5–4 cm long, leaves gray green, 1.5–3 mm wide .... *C. versicolor*

Crocus vernus (Linnaeus) Hill

PLATE 269

Crocus vernus ($2n$ = 8, 10, 12, 16, 18, 20, 22, 23) is the common European spring crocus, carpeting the alpine meadows or subalpine woodlands soon after the snow has gone. Its name means “spring.” The species (*sensu lato*) occurs in a very wide area from the Pyrenees through the Alps to western Ukraine and southwards to Sicilia and Albania where it grows from 300 to 2500 meters.

Crocus vernus in the wide concept is a very large complex which Brian Mathew in his monograph divided into two subspecies: subsp. *albiflorus* for all the small-flowered forms and subsp. *vernus* for all the large-flowered forms. This concept is now questioned by several explorers. In this book I prefer to split this complex into three species, although it is very possible that further dividing will follow when more research is carried out on the molecular level. The large-flowered subsp. *vernus* I split into *C. vernus* and *C. heuffelianus*. A large-flowered group from the northwestern part of the range I regard as a separate species, *C. heuffelianus*, with two subspecies. The small-flowered forms are treated as *C. albiflorus*.

Crocus vernus, as defined here, is the most popular species in cultivation. It is widely grown in gardens and used for winter forcing under the name "large Dutch crocuses." Some of these garden varieties produce the largest flowers and are regarded as most beautiful for planting in the gardens. The large colorful catalogs offer many named varieties in all sorts of color from pure white through pale to very dark purple or purple- and white-striped in various degrees. These plants have mainly originated from the Italian forms sometimes regarded as *C. napolitanus* or *C. neapolitanus* (later synonym). They all are of hybrid origin and I think that it would be best to follow J. Bergman who named them *C.* ×*cultorum*.

Crocus vernus in the wild is more often a woodland plant, sometimes growing at quite low elevations. At times it occurs in open mountain meadows, but not in high-elevation short alpine grassland. Flower color is extremely variable, ranging from mostly purple to striped, but there are populations with pure white flowers as well. Flowers are large, up to 5.5 cm long, and have a yellow or orange style usually well exceeding the length of the anthers.

I grow several forms originally collected in Croatia, Romania, and Austria. Most of them are raised from seed and are somewhat variable although in general dominate lilac-violet shades. Among them are some with a lilac base color surrounded by a white zone and slightly striped on the backs. I have never gotten pure white specimens.

Crocus vernus thrives in the open garden and readily sets seed. In cultivation there is a very old form distributed under the name ***C. vernus* var. *graecus***. Its flowers are somewhat smaller in size and lavender-blue with a darker margin and a deep blue blotch at the tepal base stretching down the flower tube. As it never sets seed I think that it is a sterile garden selection to which the Latinized name was applied. Most of its stocks are virus infected.

I replant my *Crocus vernus* corms once in two to three years when the clumps get too congested. I have never tried them in the greenhouse. When growing this crocus in pots, it is important to remember that the plants come from areas with a not very hot or dry summer.

Crocus heuffelianus Herbert **subsp. *heuffelianus***

PLATES 1, 270–272

Crocus heuffelianus subsp. *heuffelianus* is similar to *C. vernus* in size but is easily distinguished from the latter by the very typical dark V-shaped blotch at the tepal tips. Even white-flowered specimens have this blotch, though it is a very pale blue. Forms with a similar pattern are found occasionally in other species. I have found such specimens in *C. veluchensis* but they are easily separable by the very early rooting.

I have seen subsp. *heuffelianus* mostly in the eastern Carpathian Mountains where it is very uniform in overall appearance. I have also received similarly colored plants from the neighboring Czech Republic and I am sure that they are worth naming as a separate species. Subsp. *heuffelianus* grows primarily in clearings in mixed deciduous-coniferous mountain forests, among shrubs, and on small meadows. In lowland I came across it in dense deciduous forest of low trees and shrubs. It always grows where the soil is vernally very wet, sometimes even damp, and I have seen flowers arising above the water surface in shallow gullies brimming with melted snow. In the mountains I encountered it only on clay-based soils, but in a lowland population near Velikije Mosti in the Lvov district it grew on humus-rich sand-based soil along with *Corydalis solida* and *Alnus incana*. In mountain populations the filaments usually are of the same length as the anthers, in lowland populations the anthers are one and a half to two times longer than the filaments. Russian botanists even used this feature to separate the two distinct varieties.

I got to know *Crocus heuffelianus* long before I began my travels to the mountains. During the communist regime a favorite holiday was International Women's Day on 8 March, when it was common to give bouquets to wives, girlfriends, daughters, and female colleagues. I was a schoolboy then without much pocket money. The local people in Ukraine and the Caucasus collected huge amounts of flowers for that day, mostly wild cyclamens in the Caucasus and wild *C. heuffelianus* in the Carpathians, and made small bunches which were cheaply sold in the northern parts of the Soviet Empire. Big money could be earned on that holiday. I usually bought a small bunch of crocus flowers for my mother. Quite often some of the flowers were still attached to a young corm which had just started to form on top of the mother corm. I separated these from the bunch and planted them in a pot but never succeeded in growing a plant. It wasn't until 1975 that I collected my first *C. heuffelianus* corms in the eastern Carpathians.

In typical forms of the species, flowers are very bright purple with a well-defined darker mark at the tip of the sepals and a bright yellow or orange style, which in the type subspecies is usually at the same level as the anther tips. This feature along with the flower color distinguishes this crocus from another large-blooming relative, *C. vernus*, in which the stigmatic branches well overtop the anthers. White-flowered specimens are quite rare and there can be large populations without any white ones among slightly variable purple-colored plants. The population on the Lizja mountain pass was exceptional and consisted of approximately equal proportions of pure white forms with a beautiful slightly bluish blotch at the tip and deep purple forms, with no intermediates. A picture from this population is published in my previous book *Buried Treasures* (2007).

When Aino Paivel of the Tallinn Botanic Garden told me that she had seen some individuals with clearly pink flowers on the Srednij Vereckij mountain pass, I decided that I had to go find them. I carefully prepared for my first faraway trip to the mountains. Recalling my sad experiences with cormlets from the 8-March-holiday bouquets of my childhood, I took along many tiny clay pots in which to plant the collected corms together with a pinch of soil. I didn't find the pink form but instead I discovered, and I am sure of that, one of the best crocuses of my life—of purest white with a very deep purple blotch at the tip of the tepal. This beauty reached my garden safe and sound and started to grow and multiply. Later I gave it the cultivar name **'Carpathian Wonder'** (Plate 271). Unfortunately the original stock during one winter was completely destroyed by rodents but luckily I found a self-sown seedling of it with an almost identical color, only with a slightly thinner purple pattern. Now it is one of most searched-for crocus varieties that I have raised. In 2003 it received the Preliminary Commendation award from the Royal Horticultural Society.

Another seedling had a similarly colored flower but with a light lilac base color instead of pure white. I named it **'Dark Wonder'**. A very deep purple selection was named **'Purple Eyes'**. One time only I found a specimen with a purest white flower without any trace of blue on the segments.

Crocus heuffelianus usually increases very well by division of offsets and abundantly sets seed. It self-seeds and even has naturalized here. In my old garden it grew and bloomed in an old apple orchard for more than 20 years without any attention.

When I was growing *Crocus* ×*cultorum* varieties in my first garden, a small self-sown group of very specially colored specimens appeared. Their color resembled *C. heuffelianus* but the flowers were larger and more robust, and the rate of increase was excellent. I selected three of the most distinctly colored seedlings, which later were named **'National Park'** (Plate 272), **'Brian Duncan'**, and **'Wildlife'**. The darkest of them is 'National Park' with an almost blackish purple blotch at the tip of the outer segments extending downwards in the form of a narrow deep purple median stripe which at the base again widens into a deep purple basal blotch.

Crocus heuffelianus is a plant from comparatively wet growing conditions and it will benefit if the soil does not dry out completely during the summer months. It can be left in the ground for several years, and it feels better and blooms longer in a slight shade from neighboring trees and shrubs, but it also can be grown in full sun without problems. In the Tallinn Botanic Garden it self-seeds until it now covers the entire rock garden. It can be grown in pots but they have to be taken out of the greenhouse as soon as the weather permits. I store corms in thin plastic bags, but they can also be stored in open boxes for short periods. The best practice is to replant them after lifting as soon as possible.

Crocus heuffelianus **subsp.** ***scepusiensis*** (Rehmer & Woloszczak) Dostál

PLATE 273

A typical *Crocus heuffelianus* has a nude throat. More to the north, in Poland, it is replaced by a form with a hairy throat known under the name *C. scepusiensis*. In other features this form is almost identical with *C. heuffelianus*, except the chromosomes which are reportedly different. I prefer to regard the plants from Poland on the subspecific level as *Crocus heuffelianus* subsp. *scepusiensis*.

The style branches in subsp. *scepusiensis* are always positioned well over the anthers and in this feature it is close to *Crocus vernus*. Among seedlings of subsp. *scepusiensis* received from the Krakow Botanic Garden, Poland, most plants had pure white stigmas. Following G. Maw, who named a similarly colored form among *C. vernus* as var. *leucostigma*, I applied the same name to my selection. This form increases readily by splitting and possibly it would be better to give it a cultivar name.

In the garden subsp. *scepusiensis* requires similar conditions as the type subspecies *heuffelianus*.

Crocus albiflorus Kitaibel ex J. A. Schultes

PLATE 274

Crocus albiflorus ($2n = 8$) is a typical plant of alpine meadows where it grows in short grassland from 600 to 2500 meters and occurs in a very wide area from Spain in the Pyrenees as far as the former Yugoslavia. Alexander Leven found it growing with *C. vernus* in mixed populations in the southern French Alps on St. Bernard Pass, in the Bernese Oberland, and in the Dolomites. *Crocus albiflorus* is easily distinguishable from its cousins by smaller flowers and stigmatic branches, which usually are positioned well below the tips of anthers, although in some populations they end at the same level as the anther tips.

In flower color *Crocus albiflorus* covers the full range of variability from the purest white to deep purple with variously striped forms. The most common forms are white-colored with the outer tepal base pure white or striped purple or deep purple extending to a short deep purple "tongue" at the tepal base. Purple forms are not very common, at least among the many samples of *C. albiflorus* seeds that I have received from various sources. In 2008 I received seeds collected by Dr. Adelaide L. Stork in Switzerland in the Jura Alps near La Dole, at an elevation of approximately 1400 meters. She wrote that the seed was from a population in which striped forms dominated.

In gardens *Crocus albiflorus* is very easy. I grow it without special care on open beds. It tolerates summer rains, multiplies well by division of offsets, and readily sets seed. Because it comes from open meadows, it prefers a sunny position in the garden. It can also be grown in pots, which I set outside when the weather allows.

Crocus tommasinianus Herbert

PLATES 275–281

Crocus tommasinianus (2*n* = 16) was named in honor of the researcher of the Dalmatian flora, botanist Muzio de Tommasini. It occurs in southern Yugoslavia, southern Hungary, and northwestern Bulgaria, growing on limestone formations in deciduous woods and shady banks or among rocks from 1000 to 1500 meters where it blooms as soon as the snow melts.

Crocus tommasinianus has a slender flower bud compared with the thicker, clumsier buds of *C. vernus*. In full sun *C. tommasinianus* opens flat and starry, while the flowers of *C. vernus* and its allies have a wine-glass shape. The flower usually is pale lilac with a silvery or buff exterior and a white or pale whitish flower tube. The leaves of *C. vernus* usually are wider, too. The style branches are yellow to orange and are positioned more or less at the tips of anthers. In the wild the flower is not very variable in color; it is known mostly in pale lilac forms. White forms are rare.

Plants of *Crocus tommasinianus* in cultivation are extremely variable in color. Many of them are hybrids and some are sterile. The fertile forms naturalize in grass if it isn't too dense, among shrubs, and so forth (Plate 3). I have seen gardens where seed has scattered all over rockeries, and there among the seedlings are nice color forms. I have also seen the extreme variability of the species in the garden of the famous snowdrop specialist John Grimshaw in the United Kingdom. Grimshaw has selected many excellent forms with a pinkish color, bi-colored with white or purple markings at the tips, very contrasting in color between the outer and inner flower segments, and so forth (Plates 279 and 280).

There are several well-known garden forms as well. My favorite is the deep purple **'Ruby Giant'** (Plate 278) but, unfortunately, it is sterile and many stocks of it are virus infected. **'Pictus'** (Plate 277) is somewhat similar to *Crocus heuffelianus* in color, but the tips of its flower segments are distinctly white. **'Bobbo'** (Plate 275) is lilac with large white tips of the tepals and of more rounded shape than the typical *C. tommasinianus*. **'Eric Smith'** is silvery white and reported to form eight tepals when well established. Supposedly this cultivar is a hybrid with *C. vernus* as the pollen parent. There are also several white-colored selections. My Lithuanian friend Eugenius Dambrauskas selected a very spectacular form from open-pollinated seedlings of *C. tommasinianus*. It was later named **'Lilac Striped'** (Plate 276). The flower had wide lilac segments edged white and with inconspicuous white striping.

In 1980s I obtained seeds of *Crocus tommasinianus* from the Nikitsky Botanical Gardens in Yalta. Among the resultant seedlings appeared one with large bicolored flowers clearly showing its hybrid origin with some *C.* ×*cultorum* variety. Its sepals are distinctly silvery on the outside, but the outside and inside of the petals is bright violet. I named this seedling **'Yalta'** (Plate 281) after the place from where the seeds originated, although it was not a very good name because it was in Yalta where Josef Stalin forced Winston Churchill and Theodore Roosevelt to sign a treaty that allowed Russia to occupy the Baltic States after the end of the World War II. It is too late to change the name as the variety has become well known and is now widely grown in the Netherlands and offered in many catalogs all over the world.

The biggest problem when growing *Crocus tommasinianus* is its susceptibility to virus infection, but it is a common problem of many long-cultivated stocks. Every spring it is essential to perform careful inspection. I recommend that growers collect and sow seeds to raise their own clones which most likely will be virus-free.

Crocus tommasinianus is very easy to grow in the outside garden and to use for naturalization. It also grows well in pots and doesn't require dry and warm conditions in summer.

Crocus kosaninii Pulevič

PLATE 282

Crocus kosaninii ($2n = 14$) is a newer species named in honor of the famous Yugoslavian botanist Nedeljko Košanin who described *C. pelistericus* and *C. scardicus*. It grows in lowland oak forests and mixed scrub within a limited area in southern Serbia.

Crocus kosaninii is easily distinguishable from other species of the *C. vernus* group by its yellow throat and yellow filaments. Its flowers are lighter or darker lilac-blue and with a more or less prominent slightly darker striping. In some forms the tepal outside has a few dark stripes. The filaments and anthers are yellow; the style is divided into three short orange branches. The fibrous corm tunic is reticulated at the apex but parallel toward the base. I would not include it among the most spectacular species, but it is free flowering and readily sets seed.

Crocus kosaninii grows easily in the garden and is quite hardy. During a very severe winter, however, I almost lost my entire stock; only a few corms survived. The species does not need a hot and dry storage during the summer months.

Crocus malyi Visiani

PLATES 283 & 284

Crocus malyi ($2n = 30$) is known from a rather limited area in the Velebit Mountains in Croatia. It was named for the Austrian plant collector Karl Franz Josef Maly. This crocus grows on limestone formations from 300 to 1000 meters in short grassland and pine woodland margins. On the basis of its morphological features it is placed in series *Versicolores*, but phylogenetic studies show that it is distant from the other species in the series, and its position still is unclear.

Crocus malyi is one of the less variable species. Its flowers are pure white. The only known variation occurs in the color of the flower tube and the base of flower segments, which can have some bluish or brownish staining. The color of the style branches can vary from yellow to orange.

This species is similar to some of the white forms of *Crocus vernus* but is easily distinguishable by its yellow throat. Although it is usually described as a species without stripes on the outsides of the sepals, in some specimens I have observed a short and very fine purple middle line at the tip of the outer sepals. This line is so indistinct that it does not invalidate the descriptions of *C. malyi* as an "unstriped species." Despite the lack of variation in flower color, a few cultivars have been selected in which the distinction is mostly based on the color of the flower tube and the basal blotch, and on blooming time.

In gardens *Crocus malyi* is one of the easiest species and one of the few to not suffer even in the harshest and most unfavorable winters. I grow it outside where it readily sets seed, although I have never noticed any self-sown seedlings. It is very easy to grow in pots, but must be brought outside as soon as the weather permits. This late-blooming species is one of the last to open among the crocus species grown outside.

A specimen with very nice, nearly bicolored, large violet to purple flowers appeared at the Gothenburg Botanical Garden among a group of open-pollinated seedlings of *Crocus malyi* (Plate 284). The outer flower segments are darker, the inner slightly lighter. The throat is indistinct, in the same lilac color as the inside of the flower segments, and the filaments are the same. The flower tube is deep lilac. In everything else it looks much closer to *C. vernus* and I would have never linked it with the name of *C. malyi* had it not been raised from *C. malyi* seed. The possibility of hybridization with *C. vernus* seems quite improbable as the two species have very different chromosome numbers. Still a possibility remains that some foreign seed had fallen between the seeds of *C. malyi.*

Crocus versicolor Ker Gawler

PLATES 285 & 286

Crocus versicolor ($2n = 26$), as can be concluded from its name, is one of the most variably colored species. It occurs in southeastern France where it grows in the Alpes Maritimes and Hautes-Alpes up to Grenoble and southwards enters northwestern Italy. It can be found on limestone formations from sea level up to 1300 meters in open woodland and on grassy or rocky places.

The main color of the flower segments ranges from white to lilac and deep purple. The flower is always strikingly marked with purple lines and stripes, usually on both sides of the flower segments. Sometimes the markings are very light becoming deep purple only at the base. Usually the inner segments are less deeply marked than the outer.

In the 19th century *Crocus versicolor* was very popular in gardens and as many as 18 named varieties were grown, but since the second half of the 20th century, only **'Picturatus'** remains, known also as "Cloth of Silver." In my garden this cultivar flowered poorly and irregularly, although it increased very rapidly by splitting and never set seed. In the 1980s Michael Hoog presented me with several hundred wild-collected seeds. When these started to bloom I was pleasantly surprised by the incredible variability of the seedlings, their rich colors, the large size of flowers, and the excellent growing capacity. They were so much better than the 'Picturatus' in my garden that I destroyed my entire stock of that cultivar. I have selected more than 10 very distinctly colored seedlings; their stocks are increasing quickly and I hope that some of them will soon be available to the public.

Crocus versicolor is easy to grow in the outside garden. It suffers only in exceptionally severe winters, but I have never lost it completely. Some corms have always survived and the stock has been restored with the help of new seedlings as this crocus readily sets seed. It flowers in my garden in the second half of the spring crocus-blooming season when the weather is warmer and more favorable to pollinators, so good seed crops are usual without any additional hand-pollination. I haven't tried growing it in pots, but it is unlikely that any problems would arise. *Crocus versicolor* doesn't require a warm and dry summer rest but it does need a well-drained position in the garden.

Crocus imperati Tenore

PLATES 287 & 288

From a taxonomic perspective, *Crocus imperati* ($2n = 26$) is most closely related to *C. versicolor.* It was named after the 16th-century Italian botanist Ferrante Imperato. The species is distributed from Naples south to Calabria and on the Isle of Capri, and it grows from sea level up to 1350 meters in thin woodland and in grassy places. It has larger more extravagantly marked flowers than the more northerly growing *C. suaveolens* (formerly regarded as subspecies of *C. imperati*).

Crocus imperati is a very beautiful species and it is easy in gardens. It has the largest flowers among the Italian crocuses and as a rule is very strikingly colored. The outsides of the flowers segments usually are paler, whitish or buff toned, with more or less conspicuous deep purple stripes; the insides in most cases are bright lilac-purple, although some white forms are known as well. Some of the whites are entirely white or buff on the outside of the outer tepals though sometimes there can be a few purple stripes.

Maw (1886) described a very beautiful pinkish tinged form which he named **var. *reidii.*** Only one plant was known. According to E. A. Bowles, the original specimen was lost and new similar ones were not found. Today it is possible every now and then to obtain plants under the name var. *reidii* (Plate 288), but their flower is not "rose-pink." Furthermore, they lack the white spot at the tepal base, and the buff ground color on the outside mentioned by Mr. Reid in his letter to Peter Barr (cited by E. A. Bowles) is also absent. I have a form collected near Amalfi, in which the bright yellow throat is margined by a diffused white band before the tepal turns purple; the flower segments are narrower in that one, too.

Tony Goode of the United Kingdom notes that despite the often inclement winter weather *Crocus imperati* makes a good garden plant in well-drained soils in a sunny spot. He has seen buds encased in ice after a January snowfall opening in perfect shape a week later. In a cool settled spell the plant can be in flower for four or five weeks at a time of year when there is little color in the garden. It is easy in pots, too.

Crocus suaveolens A. Bertoloni

PLATE 289

Crocus suaveolens ($2n = 26$) has sweetly scented flowers, as can be deduced from its name. The species is distributed north of the area occupied by *C. imperati*, from Rome south to Terracina, Fondi, and Itri. It is found at elevations up to 700 meters on calcareous soils in scrub, light pine woodlands, and grassy places.

In general, the flowers are smaller than those of *Crocus imperati* but the most significant difference is the absence of a bracteole. There are small differences in the measurements between filaments and anthers and they are noted in the key. The flowers of *C. imperati* in most cases are more densely marked on the outside, but in *C. suaveolens* usually there are one long median stripe and one smaller stripe on each side of it. The style branches are somewhat paler as well. In some forms there is only one midstripe on the sepal backs.

E. A. Bowles characterized *Crocus suaveolens* as a slender, pallid likeness of *C. imperati* which "is so distinctly a poor relation of the rich and important *C. imperati* that none but those in search of modest charm and botanical interest need notice it." The cultivar **'De Jager'** is available in the trade; it has a distinctly buff-colored back of the flower segments which fades with age.

Both species are quite easy in the garden. I have grown *Crocus imperati* and *C. suaveolens* 'De Jager' since 1977 and never lost them completely, even when they sometimes suffered. Both plants increase well vegetatively and set seed copiously. Since building a greenhouse, I grow most of my stocks in pots. The plants grow very well and bloom abundantly every spring. They don't require a hot and dry summer and are very tolerant of weather conditions.

Crocus corsicus Vanucchi

PLATE 290

Crocus corsicus ($2n = 18$) is restricted to Corsica as indicated by its name. It grows in scrub and on rocky hillsides, sometimes in grassy places, at elevations from 200 to 2300 meters.

In general the flowers of *Crocus corsicus* resemble those of *C. imperati*, but are smaller and have a white center, and the leaves are fewer and narrower. *Crocus corsicus* is often among the last crocuses to flower, while *C. imperati* is usually an early flowering species. Frequently forms of *C. imperati* with a paler yellow throat are offered under the name *C. corsicus*. Perhaps those clones are of hybrid origin. If so, the corm tunics can help in correct identification. The corm tunics of *C. corsicus* are thinner and finely reticulated at the apex, while those of *C. imperati* are coarser fibrous with fibers parallel or very weakly reticulated at apex.

The flowers of *Crocus corsicus* usually are bright lilac inside, the outer flower segments are marked with dark purple stripes on a light violet or buff-toned background. Sometimes, instead of stripes, there is a wide deep purple median blotch with a somewhat feathered edge.

There is a very beautiful pure white form with a golden-shaded back on the outer segments. Brian Mathew mentions a form with black connectives on the anthers, found in 1976, a feature not previously recorded in this species.

Crocus corsicus can be confused with *C. minimus* when you move away from the trade forms. In such situations, the features of the corm tunic are most helpful in separating the species. Flowers of *C. corsicus* generally are much larger than those of *C. minimus.*

When I had just started to collect crocuses and to study the literature about them, I visited the botanical garden at the University of Latvia. At that time I had only some 30 varieties in my collection. The curator of the rock garden, Andris Orehovs, took out of the bulb frame a small pot with a flowering crocus and asked me. "What is its name?" I looked at it and with a trembling heart whispered, "*Crocus corsicus*?" Dr. Orehovs took out the label and read on it "*C. corsicus*, spont." I had passed the test and as a reward I got this pot. I think that because I had so many times reread Bowles's handbook and the photocopy of G. Maw's monograph my brains had formed an image of many species never seen at that time *in vivo.* At that moment neurons in my brain had put together the correct combination of features and provided me with the correct answer.

At present I grow *Crocus corsicus* only in the greenhouse where each spring it produces beautiful bright violet flowers with a deep purple strip on the outside. With some risk, the species can be grown in the open garden. My Lithuanian friend Eugenius Dambrauskas grows it only outside, but at his place winters are somewhat more consistent and milder.

This species readily sets seed as its blooming is in the second half of the season when bees are more active, and no additional hand-pollination is necessary. During summer it needs moderately dry conditions. It is a very tolerant species.

Crocus minimus de Candolle

PLATES 291–293

Despite its name, *Crocus minimus* ($2n$ = 24, 25, 26, 27, 28, 29, 30) does not have the smallest flowers in the genus. That honor goes to *C. cambessedesii* from Majorca.

Crocus minimus replaces *C. corsicus* in the southern part of Corsica, although there is a narrow borderline where both can overlap. *Crocus minimus* extends into Sardinia and some other smaller islands. It occurs at lower elevations from sea level up to 1500 meters, often on granite formations on rocky and sandy places in scrub or grass. This species is one of the latest crocuses to bloom in the garden after most of the other spring crocuses have stopped blooming.

Crocus minimus and *C. corsicus* are quite similar, and for a long time the two species were confused. They are easily distinguished by the size of flowers and by the corm tunics, which in *C. corsicus* are reticulated at the apex and in *C. minimus* are parallel fibrous and only occasionally joined at the apex. The flower color of *C. minimus* also is very variable.

Forms grown by Dutch nurseries most possibly come from lower elevations as they flower much earlier than many wild-collected specimens. The commercial forms have purple markings on the outer segments which merge into a deep blackish purple median blotch that covers one third of the outer segments. In most cases commercial forms are virus infected. Just recently I bought in the Netherlands 100 corms of the new cultivar *C. minimus* **'Spring Beauty'** and almost all the plants were infected with virus.

All forms from known wild origins in my collection bloom much later. The first sample I obtained came from the Institute of Ornamental Gardening in Pruhonice (Czech Republic). It was originally raised from wild-collected seeds. In my garden, it has always started to flower only when all the other crocuses have finished and never before May, but in 1980 it started blooming on 16 May. Now I have several acquisitions from the wild and they all have this late-blooming habit.

The color of wild forms of *Crocus minimus* varies widely. A form collected in Corsica at Col de Bavella by Alan Edwards and his wife has an almost entirely deep reddish purple (aubergine-toned) back on the outer flower segments with a very narrow white rim at the edges without any stripes; the inside of the tepals is a slightly lighter shade of the same reddish purple color. The inner segments are somewhat lighter. Quite unusual for such deep lilac-colored flowers is the white stigma which normally in *C. minimus* is yellow or orange. This form was found on acidic soil in an alpine meadow near a small rivulet at 1250 meters. According to Alan Edwards 80 percent of its seedlings reproduce the unique parent color. The plant, which was named **'Bavella'** (Plate 293), received a Preliminary Commendation award in 1999 and an Award of Merit in 2004. My plants of this cultivar never produced seeds, despite additional hand-pollination.

Another form, reported as collected in Sardinia, has a very special grayish purple shade covering the outside of outer segments so densely that the white background color looks more like narrow white stripes. The inner segments are more violet than purple in color (Plate 292).

All wild forms have a minute, sharply defined pure white zone at the inner base of the tepals. Such a zone is absent in commercially available forms. I recently received from Australia a form for which I have long been searching. This form of *Crocus minimus* has the purest white flowers throughout with white filaments, slightly creamy anthers, and a bright orange stigma.

In cultivation *Crocus minimus* isn't difficult; it grows well outside and if not exposed to climate extremes is hardy. Because it is very small and blooms somewhat out of the season, I prefer to plant it in pots. Neutral or slightly acidic compost is best for this species. In summer I take the pots outside and some care is needed not to overdry the corms during dormancy.

E. A. Bowles recommends propagating this species from seed as he found *Crocus minimus* to increase slowly by vegetative means. Both forms in the trade, however, are fast increasers vegetatively and have a clump-forming tendency that doubles the number of corms every year. Wild forms are much slower in this aspect, so seed sowing is necessary.

Crocus antalyensis B. Mathew

PLATES 294–296

Crocus antalyensis is one of two species in series *Flavi* without yellow in the flower segments. It was first collected in 1966 not far from Antalya by Turhan Baytop. It grows in western Turkey on rocky slopes in clay soils in light woodland or scrub and sometimes on meadows or abandoned cultivated fields at 800 to 1200 meters.

Crocus antalyensis is close to *C. flavus* subsp. *dissectus* but is easily separable from the latter by its blue color. It is the only species of this series with blue flowers, if *C. paschei* is regarded as belonging to series *Biflori*. Its bulbs lay very deep in the ground and the corm tunics have a long brown neck, in the same manner as in *C. flavus*. Although it was thought initially to be a narrow endemic of the surroundings of Antalya, it now has been found in a much wider area.

The flower color varies from pale to rather deep lilac blue. The outside of the flower segments can be nearly unmarked or yellowish buff spotted purple, veined or striped violet on a lilac or buff background. Forms from the southern part of the area usually are darker in color and more heavily marked than those from the north, although I have found fairly light violet specimens near Antalya. There is a record of a white population with the outsides flushed and flecked with blue. The filaments are pubescent and deep yellow, the anthers yellow. The stigma is divided into 6 to 12 yellow or orange branches approximately around the level of the anther tips. Near Gündoğmuş I collected a specimen with white stigmatic branches.

A new subspecies ***Crocus antalyensis*** subsp. ***striatus*** O. Erol & M. Koçyiğit from western Turkey (southwest Antalya) just recently was described by Osman Erol, Mine Koçyiğit, Levent Şik, Neriman Özhatay, and Orhan Kucuker. It is characterized by rough papery, light brown corm tunics, leaves that exceed the flower at anthesis and are recurved, distinctly striped inner perianth segments, and deeply divided pale yellow or white style. This one crocus quite well corresponds with a plant (RUDA-006) collected by me at Gündoğmuş in 2003 (Plate 296), although Gündoğmuş is situated east from Antalya. This specimen has white stigmatic branches and filaments gradually turning white at the base. In this sample, the filaments are only slightly papillose.

Sometimes, especially when *Crocus antalyensis* is not in flower, it is not easy to tell it apart from *C. flavus* subsp. *dissectus*. Several times I collected one, thinking I had the other. The corm tunics of *C. flavus* are more coarsely fibrous and its leaves are generally wider, but identical with *C. antalyensis* in subsp. *sarichinarensis*. Maybe the last could be regarded as a yellow-colored form of *C. antalyensis*. Then *C. antalyensis* would be the second species after *C. danfordiae* with coexisting blue, white (very rarely), and yellow forms.

Here it is a little tender species and with me it has sometimes suffered even when grown under protection in the greenhouse, although I lost it only when I grew it in the outside garden. After cold winters in the greenhouse it flowered poorly. In the wild its corms lie deep in the ground and so deeper pots would be advisable in cultivation. In summer they receive a dry and moderately hot rest. The pots are placed outside but under some shelter to protect them from rains. This very beautiful crocus is worth the additional care.

Crocus candidus E. D. Clarke

PLATES 297–300

The species name means "pure glossy white." I can agree with "white" in typical forms, but not with "pure glossy," especially when the flowers are not widely open. It is a plant from the northwestern part of Asiatic Turkey and is recorded from sea level up to 300 meters, but I collected it at higher elevations up to 450 meters. In the wild it grows on limestone formations in clearings in scrub and on open stony places.

I came across *Crocus candidus* when our team was searching for *Galanthus trojanus* near its *locus classicus*. There next to *C. candidus* was *C. pulchellus* (a little higher, under pine trees, grew a third species, namely, *C. flavus* subsp. *dissectus*). Both species had finished flowering but were easily separable by the width of the leaves. Nonetheless, I confirmed identification of *C. candidus* only when, in a smaller depression on a shaded side of the slope where the snow cover lingered, I found blooming specimens. Although I saw the last flowers of the season, they were quite variable in the outside coloring of the flower segments. All of them had a white background, but the color of the backs of the outer segments varied greatly from grayish or lilac-dotted, speckled or striped, and differed a little in the shape of tepals, which in the best forms were very wide and more rounded in appearance. The flower tube was almost invariably lilac or purplish brown speckled. The flower inside really was glossy white with a large deep yellow throat.

Crocus candidus has yellow filaments and anthers. The yellow or orange style is divided into six branches, positioned at the tips of anthers or slightly higher.

A very distinct feature of the species is the number of leaves, and even the flowering corms usually have only one very wide leaf. Plants with two leaves are seldom recorded, although the number of leaves can reach as many as four. *Crocus olivieri* also has wide leaves but they usually are a little narrower and, on average, more numerous. The two species are close relatives and can be cross-pollinated.

In cultivation, most plants distributed under the name of *Crocus candidus* are yellow-colored hybrids with *C. olivieri*, the true *C. candidus* having been reintroduced into gardens only recently. These hybrids are known under various names. The two main clones are **var. *subflavus***, with rounded golden yellow flowers speckled gray on the outside, and **'Little Tiger'** (Plate 300), with brown striped backs of somewhat slender flower segments and with more leaves.

E. A Bowles in his handbook described *Crocus candidus* as very variable in color. The first introduced specimens were white, but later plants were yellow and even deep orange, "some of which produce both white and yellow seedlings," thus confirming that they must be hybrids. It is not easy to understand what E. A. Bowles had in mind when writing about the deep orange color—whether it is the yellow-orange shade known in the best forms of *C. gargaricus* and *C. olivieri*, or a true deep orange color.

A few years ago crocus enthusiast Ibrahim Sözen from Turkey visited a site where only *Crocus candidus* was growing. He found two specimens with a clearly yellow flower color and named them **'Orangino'** (although in his pictures I cannot see anything orange). They grew about 10 cm apart, so couldn't be regarded as clone (Plate 299). They had a very rounded, globular appearance, with minute brown spots on the outside of the outer tepals. This discovery shows that yellow-colored forms can be found in the wild, but they are very rare.

I can grow the typical *Crocus candidus* only in a greenhouse, provided the plant gets a dry summer rest. The yellow hybrids are more tolerant and survived in my outside garden until that very harsh winter, when they perished.

Crocus boulosii Greuter

Crocus boulosii (2*n* = ?) in this chapter stands somewhat solitary as it belongs to series *Aleppici* where are joined species flowering in autumn and winter. It comes from Libya and was found by Loutfy Boulos in 1967 at an elevation of 500 meters on limestone formations at the edge of stony fields bordering on *Juniperus* and *Pistacia* woods.

In the wild *Crocus boulosii* grows in places where the annual precipitation is only 500 mm in the form of rain or snow in December and January. It blooms in midwinter when moisture is available. Its flowering usually starts only in January, so I included it between spring-blooming crocuses. The flowers are white with a grayish blue stain; the throat and filaments are yellowish or white. The anthers are yellow with a violet suffusion. The style is divided into many slender branches. The closest relatives of this species are *C. aleppicus* and *C. veneris.*

Crocus boulosii is very rarely cultivated and I haven't even searched for it, assuming that I would never be successful with it. There are reports about its cultivation in some collections. It is suitable only for pot growing with a long, dry and hot summer rest.

Little Known and New Crocuses

Crocuses have long been, and continue to be, plants of special interest because of the beautiful and variable flowers of high ornamental and garden value. More and more people are finding new plants previously unknown to science as they travel. Among these discoveries are crocuses. Nearly 20 new taxa have been discovered and described in recent decades, and the process is still going on.

Crocus boissieri Maw was encountered only once and there is only one herbarium sheet of it. Unfortunately, the type is incomplete and the sole features separating this species from other species are the extremely long filaments (1.3 cm) and short anthers. Nothing similar is known in other species. Sadly, when the locality was revisited, no crocuses were found, and the reported flowering time (June) seems incredible for such a low elevation (100 m), as that is the time when the vegetation there is sun-baked and absolutely arid, not the kind of habitat associated with blooming crocuses. It is fairly clear from the herbarium specimen that this is a distinct species which still waits for its rediscovering and more detailed studies.

In the same province where *Crocus boissieri* was reportedly found, a very interesting crocus was photographed in autumn 2008 (Plate 301). Lots of autumn-flowering *C. kotschyanus* and *C. cancellatus* grow in the area, but among them were found three specimens with large bright lilac-blue flowers and very long white anthers. The three-branched style does not exceed the tips of the anthers. The throat of the flowers seems to be pure white but in the base zone on the tepal outsides they are nicely deep lilac striped on a white background. The corm tunics are coarsely netted like those of *C. cancellatus*. When Kees Jan van Zwienen of the Netherlands photographed this crocus he assumed that it could be a natural hybrid between *C. kotschyanus* and *C. cancellatus*, but these two species are so different and taxonomically so distant that their interbreeding seems next to impossible. I didn't find any similar crocus when I revisited the area in autumn 2009, but the later finding of *C. cancellatus* subsp. *mazziaricus* with an undivided stigma and deep purple throat that made it very similar to *C. mathewii* (see p. 75) seemed to clarify the situation. Most likely they were only unusual forms of *C. cancellatus*.

Another very interesting crocus suddenly appeared among seedlings raised from wild seeds collected in Iran under the name of *Crocus caspius*. The flowers are white with purple stripes on the backs of the outer segments. Inside they are purest white with a deep orange edge surrounding the yellow throat. Most interesting are the anthers which have black connectives, and the bright red many-branched stigma which well overtops the anthers. The style in *C. caspius* is known to be three-branched only, in this seedling the style branches at the top are quite deeply subdivided; however, it is possible that it is only an unusual form of *C. caspius*. The black connectives of the anthers add to the uncertainty about its identity.

In spring 2009 a group of travelers covered a very long distance along the Mediterranean coast of Turkey. In mountains to the east from Antalya they took pictures of a very beautiful bright blue crocus from the *Crocus biflorus* group with shining black anthers (Plate 302). In this region the commonly occurring subsp. *isauricus* has anthers with black connectives in a large proportion of specimens, but usually it has four to seven leaves and they are only 0.5 to 1 mm wide. The black-anthered plants have only two or three leaves and they are much wider. This feature brings this potentially new subspecies closer to subsp. *crewei*, but the latter is distributed more to the north and is known to have white-colored flower segments.

Following their route next year I tried to find this beautiful crocus, but it was too late there. This only confirmed how important it is to be in the correct place at the correct time. The blooming period of crocuses isn't long and later their tiny leaves are hidden in grass. But instead of this blue-colored plant our team, which was north of Silifke, discovered another crocus with black anthers and white base color of flower segments. Superficially it resembles *Crocus biflorus* subsp. *crewei*, but it is growing far to the east and is much leafier—flowering plants have (5)6–7 leaves (Plates 160 & 303). It is growing at an altitude between 1500 and 1600 m on open spots with little other vegetation. This one certainly is a new subspecies waiting for description.

I collected a very beautiful member of the *Crocus biflorus* group in Iran just before the city line of Saqqez. In this part of Iran only subsp. *adamii* or subsp. *tauri* could grow but the collected sample (WHIR-163) is so unusual in color (Plate 304) that it may be something new, too. A very beautiful crocus (JRRK-075) was collected by our team east of Malatya at an altitude around 1900 m. It could be subsp. *tauri*, but again more detailed research on the molecular level is needed to check its status (Plate 305).

My first stock of *Crocus scharojanii* came from Lithuania though it was collected in the Caucasus. All inquiries about the exact location of its origin remained unanswered. Among the uniformly looking corms in spring appeared some lilac blooms. Superficially they resembled *C. kosaninii* that grows in the wild a great distance further—in southern Serbia (former Yugoslavia). If these corms were collected together with *C. scharojanii*, it can only mean that there is a not yet described spring-blooming species in the main Caucasian Mountain ridge. Unfortunately, I lost all corms of this mysterious crocus in a black frost.

On my only trip to Iran, I found a new subspecies of *Crocus speciosus* (Plates 25 and 26). Another discovery enlarged the area of *C. biflorus* subsp. *tauri* some 700 to 800 kilometers to the east and I think that the plant photographed there (Plate 176) most probably belongs to another new subspecies.

Oron Peri (Israel) in the Moab region, Jordan, found a very interesting crocus with a strong scent of *Hyacinthus orientalis*. The corm tunics closely resemble those of *C. hermoneus*, but the flower initially looked more like that of *C. aleppicus*. They are slender, white with an almost indistinguishable dark stripe along the tepal back, and the anthers are almost black, but the yellow style is many-branched as in *C. her-*

moneus. This crocus had three or five very narrow leaves less than 0.5 mm wide. Peri (2008), who found only three plants of this crocus, believes that it could be a new subspecies of *C. hermoneus* or even a new species.

I am positive that many more discoveries will be made when traveling in autumn becomes more popular. It seems quite strange that no autumn-flowering crocuses from the *Crocus biflorus* group have been found in eastern Turkey and in Iran. With the constantly growing interest in these beautiful plants, more and more new discoveries are to be expected.

⌘ Hybrids

CROCUS SPECIES ARE very conservative and hybrids between them are not common. A few hybrid groups are widely available in the trade. Most common among them are the large spring-blooming Dutch crocuses (*C.* ×*cultorum*) derived from *C. vernus*. In some cultivars *C. tommasinianus* is involved as well.

Another very popular group has the trade name *Crocus chrysanthus* hybrids, but it encompasses many hybrids between *C. chrysanthus* and *C. biflorus sensu lato* as well as hybrids between different subspecies of *C. biflorus*, so it would be better to give this group a hybrid name. I would like to propose *C.* ×*annulatus* but unfortunately this epithet was already used in 1820.

A third popular hybrid is the common large yellow garden crocus, a cross between *Crocus flavus* and *C. angustifolius*. Leonid Bondarenko and I created a new group of hybrids between *C. reticulatus* and *C. angustifolius*, named here as *C.* ×*leonidii*. A few of these have found their way to the world bulb market.

Crocus* ×*jessoppiae was described in 1924 by E. A. Bowles who assumed that it was a hybrid between a white form of *C. chrysanthus* and either *C. candidus* or a species with strongly netted corm tunics. Brian Mathew thinks that one of the parents could be *C. reticulatus*. The flowers are somewhat similar to those of *C. pestalozzae*, but larger and with stronger blue markings at the base of the outer segments, the throat is yellow, the anthers are yellow, and the stigmatic branches are yellow-orange. The corm tunic is similar to that of *C. candidus*, but the leaves are narrower and the stigma divides into only three branches. This hybrid appeared between some seedlings and offsets that Bowles had given to a neighbor, Miss Euphemia Jessopp. The plant is very easy to grow and multiplies quickly in the garden.

Several hybrids can be found in the wild where the areas of two species overlap. Some of them have been given species names:

Crocus ×*bornmuelleri*, a hybrid between *C. chrysanthus* and *C. biflorus*.
Crocus ×*fritschii*, a hybrid between *C. vernus* and *C. albiflorus*.
Crocus ×*koritnicus*, a hybrid between *C. chrysanthus* and *C. biflorus* subsp. *weldenii*.
Crocus ×*nubigenioides*, a hybrid between *C. chrysanthus* and *C. biflorus* subsp. *adamii*.
Crocus ×*paulineae*, a hybrid between *C. abantensis* and *C. ancyrensis*. This cross was repeated by me in my nursery using *C. ancyrensis* as seed parent. It resulted in very dramatically colored seedling (Plate 306).
Crocus ×*petrovicii*, a hybrid between *C. chrysanthus* and *C. biflorus* subsp. *alexandri*.
Crocus scharojanii var. *flavus*, a hybrid between *C. vallicola* and *C. scharojanii*.
Crocus vallicola var. *intermedia*, a hybrid between *C. vallicola* and *C. scharojanii*.

Several additional hybrids were found in the wild but have not been given species names:

Crocus atticus × *C. veluchensis*, found in the Balkans.
Crocus boryi × *C. tournefortii*, found on eastern Crete, Greece.
Crocus chrysanthus × *C. biflorus* subsp. *isauricus*. I personally saw these on Gembos yaila near Akseki, where both species grow side by side and widely hybridize, making plants of the most unusual color combinations (Plate 307). It is a quite unique place where almost every other plant seems to be of hybrid origin. Unfortunately this "Natural Crocus Breeding Station" soon will disappear to make way for houses. The land is divided and building is started, so it is the last time to try to catch this beauty.
Crocus chrysanthus × *C. biflorus* subsp. *pulchricolor*. I personally saw this on the Ulu-Dag, where the two species grow side by side.
Crocus corsicus × *C. minimus*, found in southern Corsica.
Crocus dalmaticus × *C. atticus*, found in the Balkans.
Crocus mathewii × *C. pallasii*, found in southwestern Turkey.
Crocus speciosus × *C. pulchellus*, found in Turkey. This combination has been repeated both ways in the garden and several cultivars have been selected from it.

Several successful combinations have been made by gardeners, and in some cases very nice hybrids were produced:

Crocus atticus subsp. *sublimis* × *C. gargaricus*, reported by Dirch Schnabel, Germany.
Crocus candidus var. *subflavus*, a hybrid between *Crocus candidus* and *C. olivieri*.
Crocus gilanicus × *C. autranii* was crossed in the Gothenburg Botanical Garden.
Crocus 'Golden Yellow', an old garden hybrid between *C. flavus* and *C. angustifolius*.
Crocus ×*gotoburgensis*, a hybrid between *C. scardicus* and *C. pelistericus* raised in the Gothenburg Botanical Garden.
Crocus hadriaticus × *C. sativus* var. *cashmerianus*.
Crocus korolkowii × *C. michelsonii*.
Crocus kotschyanus × *C. ochroleucus*, a hybrid offered in the trade.
Crocus leichtlinii × *C. kerndorffiorum*.
Crocus 'Rainbow Gold', a hybrid between *Crocus cvijicii* and *C. veluchensis* made by Dirch Schnabel in Germany.
Crocus sieberi × an unknown species, possibly *C. cvijicii*, resulting in a yellow-flowered hybrid.
Crocus ×*stellaris*, an old garden hybrid between *C. flavus* and *C. angustifolius*.

Possibly there are more hybrids raised but at present I couldn't find any information about them.

⌘ Lists of Crocuses by Growing Conditions in Cultivation

It is not easy to group crocuses by growing conditions as these vary greatly from place to place according to the climatic zone in which the garden is situated. I myself divide all crocuses into three basic groups, but many of the species that can grow here in Latvia only under cover, in western Europe can be grown in the open garden as well. Therefore, I have further subdivided some of my basic groups to accommodate these differences.

Subspecies are not listed here when their growing conditions are similar to those of the species. Some species are included in more than one group when I'm uncertain about their exact growing requirements. The local conditions in your garden can also vary, as can the conditions between populations of widespread species. Despite these limitations, the following grouping is a starting point for those gardeners who develop a deeper interest in crocuses.

Group A

Needs hot and dry summer conditions. Suitable only for growing under cover.

- *C. alatavicus*
- *C. aleppicus*
- *C. asumaniae*
- *C. biflorus* (many of Asian subspecies)
- *C. boulosii*
- *C. cancellatus*
- *C. candidus* (?)
- *C. carpetanus*
- *C. cartwrightianus*
- *C. cyprius*
- *C. graveolens*
- *C. hartmannianus*
- *C. hermoneus*
- *C. hyemalis*
- *C. karduchorum*
- *C. leichtlinii*
- *C. mathewii*
- *C. michelsonii*
- *C. moabiticus*
- *C. naqabensis*
- *C. nerimaniae*
- *C. nevadensis*
- *C. oreocreticus*
- *C. pallasii*
- *C. sativus*
- *C. sieberi*
- *C. sieheanus*
- *C. thomasii*
- *C. veneris*
- *C. vitellinus*
- *C. wattiorum*

Group B

Tolerates some summer moisture. Needs a drier rest to thrive.

- *C. aerius*
- *C. ancyrensis*
- *C. antalyensis*
- *C. atticus*
- *C. biflorus* (European subspecies and subsp. *pulchricolor* and subsp. *tauri* from Turkey)
- *C. cancellatus* (a few forms)
- *C. candidus* (?)
- *C. chrysanthus*
- *C. danfordiae*
- *C. fleischeri*
- *C. korolkowii*
- *C. kotschyanus* (subsp. *cappadocicus* and subsp. *hakkariensis*)
- *C. olivieri*
- *C. robertianus*

Subgroup BB

Tolerates some summer moisture. Must be grown in pots under cover in colder climates.

- *C. adanensis*
- *C. aerius*
- *C. almehensis*
- *C. antalyensis*
- *C. biflorus* (many of the Asian subspecies)
- *C. boryi*
- *C. cambessedesii*
- *C. cancellatus* (most forms)
- *C. candidus*
- *C. caspius*
- *C. flavus* (not type subsp.)
- *C. goulimyi*
- *C. hadriaticus*
- *C. hittiticus*
- *C. kerndorffiorum*
- *C. leichtlinii*
- *C. ligusticus*
- *C. longiflorus*
- *C. melantherus*
- *C. michelsonii*
- *C. niveus*
- *C. paschei*
- *C. pestalozzae*
- *C. reticulatus* (southern forms)
- *C. serotinus*
- *C. sieberi*
- *C. suworowianus*
- *C. tournefortii*

Group C

Tolerates summer moisture. Species marked with an asterisk (*) are better grown in pots kept under cover during the winter for they could not be sufficiently hardy in Latvia or areas with a similar climate.

- *C. albiflorus*
- *C. angustifolius*
- *C. baytopiorum**
- *C. corsicus*
- *C. dalmaticus*
- *C. etruscus*
- *C. flavus* subsp. *flavus*
- *C. gilanicus**
- *C. heuffelianus*
- *C. imperati*
- *C. kosaninii*
- *C. kotschyanus* subsp. *kotschyanus*
- *C. laevigatus**
- *C. malyi*
- *C. minimus*
- *C. nudiflorus*
- *C. ochroleucus**
- *C. pulchellus*
- *C. reticulatus* (northern forms)
- *C. rujanensis*
- *C. speciosus*
- *C. suaveolens*
- *C. tauricus*
- *C. tommasinianus*
- *C. versicolor*
- *C. vernus*

Subgroup CC

Prefers not to be dried out completely at any time. Species marked with an asterisk (*) are better grown in pots kept under cover during the winter for they are not sufficiently hardy in Latvia or areas with a similar climate.

- *C. abantensis*
- *C. autranii**
- *C. banaticus**
- *C. cvijicii*
- *C. gargaricus*
- *C. herbertii*
- *C. vallicola*
- *C. veluchensis*

Subgroup CCC

Needs moist conditions the whole summer. Better grown in pots so the moisture level can be constantly kept under control.

- *C. pelistericus*
- *C. scardicus*
- *C. scharojanii*

⌘ Glossary

Acuminate Tapering into a narrow point.
Acute With a sharp point.
Annuli Rings of membranous tunics forming the lower portion of the corm tunic in certain species.
Anther The upper part of the stamen producing the pollens.
Apical At the tip or apex.
Basal spathe A tubular membrane springing from the summit of the corm and wrapping the flower stalk and ovary in certain species.
Basal tunic A separate small portion of tunic at the base of the corm.
Bract, bracteole A leaflike organ associated with a flower or an inflorescence.
Canaliculate Longitudinally channeled.
Capsule The seed vessel.
Carinate Keeled.
Caruncle A fleshy protuberance on the seed near the point where it was attached.
Ciliate/ciliolate Fringed with fine/minute hairs.
Clavate Club-shaped.
Claw Narrowed base of a tepal.
Clone A vegetatively propagated plant, all offspring are identical.
Contractile Able to shorten and pull as in contractile roots.
Coriaceous Tough, leathery.
Corm A solid bulblike underground stem.
Corm tunic Fibrous or membranous wrapping of the corm, renewed annually.
Cultivar A variant which is considered to be distinct for horticultural purposes
Diphyllous Two-leaved, used for the proper spathes (which see).
Extrorse Having anthers opening on the outer side.
Fibrous Used for corm tunics with parallel fibers.
Filament The stalk of an anther.
Glabrous Smooth, without hairs.
Glaucous With a waxy, blue-gray bloom.
Hysteranthous Used of leaves which are produced later than the flowers.
Inflorescence The arrangement of the flowering part of the plant.
Introrse Having anthers opening on the inner side.
Keel A thickened central ridge on the under side of the leaf.
Lamina Expanded portion of the leaf.
Locus classicus The classical location from where a taxon is described.
Membranous A skin of homogenous tissue forming the tunic of some species, thin and often semi-transparent.
Monophyllous One-leaved, used of the proper spathe (which see).

Multifid Divided into many parts.

Obovate Like ovate but reversed, with the broadest end nearer the apex.

Obtrullate Roughly obovate in outline but angular, like a bricklayer's trowel in reverse.

Ovary The part of flower that contains the ovules, which develop into seeds after fertilization.

Ovate Shaped like a hen's egg, broadest near the base.

Papillose Rough. Furnished with minute hairlike protuberances (bearing tiny projections.).

Perianth The outer, non-reproductive parts of flower arranged into two whorls.

Perianth tube The slender tubular portion of flower between ovary and segments, serving instead of a stalk.

Petal A term used for perianth segment of inner whorl.

Pistil The female organ consisting of ovary, style, and stigma.

Proper spathe One or two spathes springing from just below the ovary.

Pubescent Covered with fine, short hairs.

Reticulate Netted.

Scape The flower stalk below the ovary.

Segments (flower) The divisions of the perianth, arranged in an outer and inner series.

Sepal A term used for perianth segment of outer whorl.

Sheathing leaves/scales A strong wrapping of three to five tough and fleshy tubular leaves arising from the inner wrapping of the corm, and enclosing the leaves and flowers until they reach the ground level.

Spathe A thin semitransparent bract that encloses a flower before flowering.

Stamen The male organ consisting of anther and filament.

Stigma Upper part of the pistil which receives the pollen.

Stigmatic branches Three or more divisions into which the upper part of the crocus pistil is divided.

Stoloniferous Forming stolons, underground stems that produce a new corm at the apex.

Synanthous Used of leaves which are produced with the flowers.

Taxon An unspecified unit of classification.

Tepal A term used for perianth segment.

Throat The funnel-shaped orifice of the flower where the perianth tube and perianth segments join.

Toothed With small pointed projections along the margin.

Trifid Divided into three parts.

Tunic See corm tunic.

Xerophytic Growing in very dry conditions.

Yaila A high-mountain meadow; flat or with an uneven or rugged relief sometimes caused by karst processes.

⌘ Selected Bibliography

Al-Eisawi, D. 2001. Two new species of Iridaceae, *Crocus naqabensis* and *Romulea petraea* from Jordan. *Arab Gulf Journal of Scientific Research* 19: 167–169.

Amatniece, V. 1975. Krokusu pavairošanās koeficienta atkarība no audzēšanas veida *Daiļdārzniecība* 10: 75–79. Rīga: Zinātne. (In Latvian).

Avrorin, N. A., ed. 1977. *Ornamental Plants for Open Garden in USSR. Monocotyledons*, vol. 1. Leningrad: Nauka. (In Russian)

Baker, J. G. 1873. Review of the known species of Crocus. *Gard. Chron.*: 107, and etc.

Barker, C. 2008. *World Checklist of Iridaceae*. The Board of Trustees of the Royal Botanic Gardens, Kew. Published on the Internet; http://www.kew.org/wcsp.

Bird, P. 1999. *Crocus pelistericus. Crocus Group Bulletin* 27.

Bowles, E. A. 1914. *My Garden in Spring*. Reprint. Portland, Oregon: Timber Press, 1997.

Bowles, E. A. 1915. *My Garden in Autumn and Winter*. Reprint. Portland, Oregon: Timber Press, 1998.

Bowles, E. A. 1952. *A Handbook of Crocus and Colchicum for Gardeners*. Rev. ed. London: Bodley Head.

Brighton, C. A., B. Mathew, and C. J. Marchant. 1973. Chromosome counts in the genus *Crocus. Kew Bulletin* 28: 451–464.

Brighton, C. A., C. J. Scarlett, and B. Mathew. 1980. Cytological studies and origins of some *Crocus* cultivars. *Petaloid Monocotyledons*. Linnean Soc. London., Academic Press. p. 139–159.

Cobb, R. 1969. Some notes on the cultivation of *Crocus* species. *Journal of the Royal Horticultural Society* 94: 445–448.

Davis, P. H., ed. 1984. *Flora of Turkey and the East Aegean Islands*. Vol. 8. Edinburgh: University Press.

Edwards, A. 1996. *Crocus cvijicii* and *Crocus pelistericus* in Greece. *Crocus Group Bulletin* 24.

Edwards, A. 1998. A crocus foray in the Pindus. *The Alpine Gardener* 66: 361–364.

Edwards, A. 1999. Some thoughts on *Crocus sieberi*. *Crocus Group Newsletter*. Spring.

Edwards, A. 2000. *Crocus minimus* DC. 'Bavella', P.C. Iridaceae. *The Alpine Gardener* 68: 241–243.

Edwards, A. 2007. *Crocus* 'Snow Leopard', P.C. Iridaceae. *The Alpine Gardener* 75: 500–501.

Erol, O., M. Koçyiğit, L. Şik, N. Özhatay, and O. Küçüher. 2010. *Crocus antalyensis* subsp. *striatus* subsp. nov. (Iridaceae) from southwest Anatolia. *Nordic Journal of Botany*, 2010. Early view Internet publication.

Erol, O., and O. Küçüher. 2005. The crocus of Istanbul. *The Plantsman*, n.s., 4: 168–169.

Fedtschenko, B. A., et al. 1932. *Flora Turkmenii (Flora Turcomanica)*. Vol. 1 part 2. Leningrad: Academiae Scientiarum URSS. (In Russian)

Fedorov, An. A., ed. 1979. *Flora Partis Europaeae URSS*. Vol. 4. Leningrad: Nauka. (In Russian)

Feinbrun, N., and A. Shmida. 1977. A new review of the genus *Crocus* in Israel and neighbouring countries. *Israel Journal of Botany* 26: 172–189.

Goode, T. 2003. Crocuses in the garden. *The Alpine Gardener* 71: 137–143.
Grey-Wilson, C., and B. Mathew. 1981. *Bulbs. The Bulbous Plants of Europe and Their Allies*. London: Collins.
Grossheim, A. A. 1940. *Flora Caucasica*. Vol. 2, 2d ed. Baku, Azerbaijan: AzFAN. (In Russian)
Güner, A., et al., eds. 2000. *Flora of Turkey and the East Aegean Islands*. Vol. 11. Edinburgh: University Press.
Haworth, A. H. 1809. On the cultivation of crocuses with a short account of the different species known at present. *Transactions of the Horticultural Society of London* 1: 122–139.
Huber, Th. *Der Krokus. Kultivare und Species*. Unpublished.
Jacobsen, N., J. Van Scheepen, and M. Orgaard. 1997. The *Crocus chrysanthus–biflorus* cultivars. *The Plantsman*, n.s., 4: 6–38.
Kapinos, G. J. 1965. *Biology of development of bulbous and cormous plants on Apscheron*. Baku, Azerbaijan: Academy of Science of Azerbeijan. (In Russian)
Keeble, S. 1999. *Crocus* in Cyprus. *Crocus Group Newsletter*. Spring.
Kerndorff, H. 1993. Two new taxa in Turkish *Crocus* (Iridaceae). *Herbertia* 49: 76–86.
Kerndorff, H. 1994–1995. Notes on *Crocus* (Iridaceae) in Syria and Jordan. *Herbertia* 50: 68–81.
Kerndorff, H., and E. Pasche. 1994. *Crocus mathewii*. A new autumn-flowering crocus from Turkey. *The Plantsman*, n.s., 1: 102–106.
Kerndorff, H., and E. Pasche. 1996. Crocuses from Turkey to Jordan. *Quarterly Bulletin of the Alpine Garden Society* 64: 296–312, 459–467.
Kerndorff, H., and E. Pasche. 1997. Zwei bemerkenswerte Taxa des *Crocus biflorus*-Komplexes (Iridaceae) aus der Nordostturkei. *Linzer Biologische Beiträge* 29: 591–600.
Kerndorff, H., and E. Pasche. 1998. On the type locality of *Crocus boissieri* (Iridaceae). *The Plantsman*, n.s., 5: 12–14.
Kerndorff, H., and E. Pasche. 2003. *Crocus biflorus* in Anatolia. *The Plantsman*, n.s., 2: 77–89.
Kerndorff, H., and E. Pasche. 2004. *Crocus biflorus* in Anatolia. Part Two. *The Plantsman*, n.s., 3: 201–215.
Kerndorff, H., and E. Pasche. 2004. Two new taxa of the *Crocus biflorus* Aggregate (Liliiflorae, Iridaceae) from Turkey. *Linzer Biologische Beiträge* 36: 5–10.
Kerndorff, H., Pasche, E. 2006. *Crocus biflorus* (Liliiflorae, Iridaceae) in Anatolia. Part Three. *Linzer Biologische Beiträge* 38: 165–187.
Ketzkhoveli, N. N., and R. I. Gagnidze, eds. 1971–2001. *Flora of Georgia*. Vols. 1–13. Metsniereba, Tbilisi. (In Georgian)
Kolakovski, A. A. 1986. *Flora of Abchasia*. Vol. 4. Metsniereba, Tbilisi. (In Russian).
Koltsova, A. S. 1972. The biology of wild *Crocus* species under natural growth conditions. *Proceedings of the State Nikita Botanical Gardens* 59: 79–90. (In Russian)
Koltsova, A. S. 1973. *Morphogenesis of wild and cultivated species of Genus Crocus*. Moscow: Moscow State University. (In Russian)
Koltsova, A. S. 1976. Morphogenesis and duration of organogenesis stages in *Crocus* species in culture. *Proceedings of the State Nikita Botanical Gardens* 68: 83–97. (In Russian)

Komarov, V. L., ed. 1935. *Flora SSSR*. Vol. 4. Leningrad: Editio Academiae Scientarum URSS. (In Russian)

Kos, J. 1948. About two forms of *Crocus vallicola*. *Bot. J. Acad. Sc. USSR*. 33: 376–378. (In Russian)

Kudrjaschev, S. N., ed. 1941. *Flora Uzbekistanica*. Vol. 1. Taschkent: Editio Sectionis Uzbekistanicae Academiae Scientiarum URSS. (In Russian)

Leeds, R. 2000. *Early Bulbs*. Portland, Oregon: Timber Press; Newton Abbot, Devon: David & Charles.

Maroofi, H. 2002. Notes on the flora of Kurdistan province, Iran. *Iranian Journal of Botany* 9: 233–234.

Mathew, B. 1977. *Crocus sativus* and its allies (Iridaceae). *Plant. Syst. Evol.* 128: 89–103.

Mathew, B. 1982. *The Crocus. A Revision of the Genus Crocus (Iridaceae)*. London: B. T. Batsford.

Mathew, B. 1983. The Greek species of *Crocus* (Iridaceae), a taxonomic survey. *Ann. Musei Goulandris* 6: 63–86.

Mathew, B. 1985. *Crocus hadriaticus* and its variants. *Kew Magazine* 2: 309–311.

Mathew, B. 1987. *The Smaller Bulbs*. London: B. T. Batsford.

Mathew, B. 1995. An interesting new autumn-flowering *Crocus* from Turkey. *The Plantsman*, n.s., 5: 10–12.

Mathew, B. 1998. White forms of *Crocus serotinus* subsp. *salzmannii*. *The Plantsman*, n.s., 2: 182–184.

Mathew, B. 2002. Crocus up-date. *The Plantsman*, n.s., 1: 44–56; 2: 93–102.

Mathew, B., and C. A. Brighton. 1977. Four Central Asian *Crocus* species. *Iranian Journal of Botany* 1: 123–135.

Mathew, B., G. Pettersen, and O. Seberg. 2009. A reassessment of *Crocus* based on molecular analysis. *The Plantsman*, n.s., 8: 50–57.

Maw, G. 1886. *A Monograph of the Genus Crocus*. London: Dulau.

Meikle, R. D. 1985. *Flora of Cyprus*. Vol. 2. Kew: Royal Botanic Gardens.

Ovcinnikov, P. N., ed. 1963. *Flora of Tajikistan*. Vol. 2. Moscow: Editio Academiae Scientiarum URSS. (In Russian)

Pasche, E. 1993. A new *Crocus* (Iridaceae) from Turkey. *Herbertia*. 49: 67–75.

Pasche, E., and H. Kerndorff. 1999. A new natural hybrid in the genus *Crocus* (Iridaceae). *The Plantsman*, n.s., 6: 43–45.

Peeters, J. M. M., ed. 1995. *Ziekten en afwijkingen bij bolgewassen*. Deel 2: 35–48. Lisse: Informatie en Kennis Centrum Landbouw; Laboratorium voor Bloembollenonderzoek. (In Dutch)

Peri, O. 2006. *Crocus* in Israel. *Crocus Group Bulletin* 34.

Peri, O. 2008. *Crocus moabiticus*. *Crocus Group Bulletin* 36.

Petersen, G., O. Seberg, S. Thorsøe, T. Jørgensen, and B. Mathew. 2008. A phylogeny of the genus *Crocus* (Iridaceae) based on sequence data from five plastid regions. *Taxon* 57: 487–499.

Petrova, E. 1975. Pruhonicky sortiment krokusu v letech 1970–1974. *Acta Pruhoniciana* 33. (In Czech)

Phillips, R., and M. Rix. 1989. *Bulbs*. Rev. ed. London: Pan Books.

Pilipenko, V. V., V. V. Skripchinskyi, and V. G. Tanfiliev, eds. 1979. *Wild Plants of Stavropol District*. Vol. 2. Stavropol, Russia. (In Russian)

Pulevič, V. 1978. *Crocus thomasii* Ten. i *Crocus pallasii* Goldb. u Flori Jugoslavije. *Glas. Republ. Zavoda Zašt. Prirode—Prirodnjačkog Muzeja Titograd* 11: 133–138. (In Serbian)

Randjelovič, N., D. A. Hill, V. Stamenkocič, and V. Randjelovič. 1990. A new species of *Crocus* from Yugoslavia. *Botanical Magazine (Kew Magazine)* 7: 182–186.

Richards, J. 2006. South Wales show, February 12. *The Alpine Gardener* 74: 74–77.

Rolfe, R. 2000. *Crocus* ×*gotoburgensis* R. Rolfe 'Ember', P.C. *The Alpine Gardener* 68: 228–230.

Royal General Bulb Growers Association. 1975. *Classified List and International Register of Hyacinths and Other Bulbous and Tuberous Rooted Plants.* 3rd ed. Hague, Netherlands: Koninklijke Algemene Verenignig voor Bloembollencultuur (Royal Bulb Growers Association).

Royal General Bulb Growers Association. 1991. *International Checklist for Hyacinths and Miscellaneous Bulbs.* Hillegom, Netherlands: Koninklijke Algemene Verenignig voor Bloembollencultuur (Royal Bulb Growers Association).

Rukšāns, J. 1981. *Crocuses.* Riga: Avots. (In Latvian).

Rukšāns, J. 2007. *Buried Treasures.* Portland, Oregon: Timber Press.

Schenk, P. K. 1970. Root rot in *Crocus. Neth. J. Pl. Path.* 76: 159–164.

Sheasby, P. 2007. *Bulbous Plants of Turkey and Iran.* Alpine Garden Society.

Stephens, D. 1996. *Crocus* notes. *Crocus Group Bulletin* 24.

Stephens, D. 1998. *Crocus desiderata. The Alpine Gardener* 66: 353–360.

Stephens, D. 1999. *Crocus medius. Crocus Group Newsletter.* Spring.

Stephens, D. 1999. *Crocus carpetanus. Crocus Group Bulletin* 27.

Stephens, D. 2002. *Crocus* of Greece—central and eastern Macedonia and western Thrace. *Crocus Group Newsletter.* Summer.

Stephens, D. 2006. *Crocus* from seed. *Crocus Group Bulletin* 34.

Takhtajan, A., ed. 2006. *Caucasian Flora Conspectus.* Vol. 2. Saint Petersburg University Press (In Russian).

Tutin, T. G., et al., eds. 1980. *Flora Europaea.* Cambridge: Cambridge University Press.

Watt, P., and P. Watt. 2006. November in southwest Anatolia. *Crocus Group Bulletin* 34.

Watt, P., and P. Watt. 2008. Autumn in Laconia. *Crocus Group Bulletin* 36.

Wendelbo, P., and B. Mathew. 1975. Iridaceae. In *Flora Iranica*, no. 112, ed. K. H. Rechinger. Graz, Austria: Akademische Druck und Verlagsanstalt.

Wulff, E. W. 1929. *Flora Taurica.* Vol. 1. Fasc. 2. Leningrad: Nikitsky Botanical Gardens. (In Russian)

Yüzbasioglu, S., and Ö. Varol. 2004. A new autumn-flowering *Crocus* from SW Turkey. *The Plantsman*, n.s., 3: 104–106.

Conversion Tables

MILLIMETERS	INCHES
1 mm	3/64 in.
3 mm	1/8 in.
5 mm	3/16 in.
6 mm	1/4 in.
8 mm	5/16 in.
10 mm	3/8 in.
11 mm	7/16 in.
12 mm	1/2 in.
15 mm	5/8 in.
20 mm	3/4 in.

CENTIMETERS	INCHES
1 cm	1/2 in.
1.5 cm	5/8 in.
2 cm	3/4 in.
2.5 cm	1 in.
3 cm	1 1/8 in.
3.5 cm	1 1/4 in.
4 cm	1 1/2 in.
5 cm	2 in.
6 cm	2 3/8 in.
7 cm	2 3/4 in.
8 cm	3 in.
9 cm	3 1/2 in.
10 cm	4 in.
15 cm	6 in.
20 cm	8 in.
23 cm	9 in.
25 cm	10 in.
30 cm	12 in.
40 cm	15 in.
47 cm.	18 1/2 in.
50 cm	20 in.
60 cm	24 in.
70-cm	28-in.

METERS	FEET
1 m	3.3 ft.
30 m	100 ft.
50 m	165 ft.
100 m	330 ft.
200 m	660 ft.
300 m	990 ft.
400 m	1320 ft.
500 m	1650 ft.
600 m	1980 ft.
700 m	2310 ft.
800 m	2640 ft.
900 m	2970 ft.
1000 m	3300 ft.
2000 m	6600 ft.
3000 m	9900 ft.

KILOMETERS	MILES
1 km	0.6 mile
17 km	10 miles
30 km	19 miles
40 km	25 miles
700 km	435 miles
800 km	498 miles

°CELSIUS	°FAHRENHEIT
–47°C	–53°F
–35°C	–31°F
–27°C	–17°F
–25°C	–13°F
–15°C	5°F
–2°C	28°F
10°C	50°F
17°C	63°F
21°C	70°F
25°C	77°F
30°C	95°F
43.5°C	110°F

GRAMS/SQUARE METER
30 g/m^2
40 g/m^2
70 g/m^2
100 g/m^2
200 g/m^2

OUNCES/SQUARE YARD
3/4 oz/yd^2
1 oz/yd^2
2 oz/yd^2
3 oz/yd^2
6 oz/yd^2

MISCELLANEOUS

2 grams/liter
1/2 teaspoon/quart
100 grams/cubic meter
2 3/4 ounces/cubic yard
3 kilograms/cubic meter
5 pounds/cubic yard

⌘ Index

Accepted *Crocus* species names are in **bold** type; cultivars, synonyms, and other plant names are in regular type (keys are not included). **Bold** numbers indicate main entry pages.